GREENHOUSE EFFECT AND SEA LEVEL RISE

GREENHOUSE EFFECT AND SEA LEVEL RISE

A Challenge for This Generation

Edited by

Michael C. Barth
ICF Incorporated

James G. Titus
*Environmental
Protection Agency*

Foreword by
William D. Ruckelshaus
Environmental Protection Agency

 VAN NOSTRAND REINHOLD COMPANY

The studies upon which the papers in this book are based were largely financed by the U.S. Environmental Protection Agency in the interest of enhancing our understanding of important environmental issues. EPA does not necessarily endorse or authorize any of the findings in this book. The views expressed in this volume are solely those of its authors and should not be deemed to constitute the official views or policy of the U.S. government.

Manufactured in the United States of America.

Published by Van Nostrand Reinhold Company Inc.
135 West 50th Street
New York, New York 10020

Van Nostrand Reinhold Company Limited
Molly Millars Lane
Wokingham, Berkshire RG11 2PY, England

Van Nostrand Reinhold
480 Latrobe Street
Melbourne, Victoria 3000, Australia

Macmillan of Canada
Division of Gage Publishing Limited
164 Commander Boulevard
Agincourt, Ontario MIS 3C7, Canada

15   14   13   12   11   10   9   8   7   6   5   4   3   2

Library of Congress Cataloging in Publication Data
Main entry under title:
Greenhouse effect and sea level rise.
    Includes bibliographical references and index.
    1. Sea level. 2. Greenhouse effect, Atmospheric. 3. Coast changes.
I. Barth, Michael C.   II. Titus, James G.
GC89.G74   1984   363.3'492   84-7207
ISBN 0-442-20991-6

Contents

Foreword

Winston Churchill, whose output as a writer was so prodigious that one might reasonably wonder whether he was paid by the page, was himself put off by long works. As prime minister of Great Britain, he once pushed away a report from a junior minister, observing, "This paper, by its very length, defends itself against the risk of being read."

No doubt some laymen will glance at the study that follows and be deterred from further consideration by the fact that the study is a scientific one. If that is a hurdle for you, I hope you will join me in getting over it, because, if you do, you will find that the authors have quite a story to tell.

That story is about change—the possibility that the climatic patterns of the world are in a transition to warmer weather that could lead to a rise in the sea level. You may not have thought much about the sea level previously; it was something we took for granted. But since we have taken it as a given for so long, the adjustments we may have to make will be profound. When you stop to think about all the areas of our lives that could be affected by climatic change, you will be amazed: we have planned our cities, developed our manufacturing techniques, and chosen our environmental protection strategies on the assumption of a stable sea level.

And even that is not the end of the story. Just as the potential effects of sea level rise will be societal, so the decisions whether to anticipate and how to respond to the new conditions will need to be made in large part by our governmental institutions. While the magnitude of the challenge is much greater than most faced by governments in the past, it is perhaps representative of many that we will face in the future. The question is, do we have the will to begin to face these questions today?

The question is a serious one, because we will find out whether a free, democratic society can respond to pressing needs in time to have maximum effect, while nonetheless not also changing its essential character. Bismarck's view that "God looks after fools, drunkards, and the United States of America" reflects his observation that our government has always functioned well, not necessarily by anticipating crises of great magnitude but by waiting for what political scientists call "an action-forcing event." Our system of government has traditionally been biased toward a sort of institutional inertia, which eventually is broken by development of a massive consensus that sweeps through remaining barriers and ensures that the policies finally adopted will have lasting constituencies. The problem, of course, is that in our ultimate haste, we may not give adequate attention to all the options; Charles Haar calls this phenomenon "the catastrophe theory of planning."

Whether we can continue in such a manner is a subject open to question. For one thing, America's predominant place in the world requires that we act first on many questions; we can no longer depend on having the benefit of watching other Western democracies, then incorporating their experiences into our own. Another new development is that our representative institutions have become so responsive to so many groups that it is often hard to get a great deal done. Just imagine the government trying to set up the TVA today, in view of the legal, bureaucratic, and congressional barriers it would face!

Perhaps the most significant difference is in the nature of the issues themselves. In many areas, including the environment, government must act to anticipate problems in order to avert potentially serious consequences. In addition, it must act on the basis of knowledge that is often technical in nature (and thus hard to communicate and easy to distort—factors of no small consequence in democracies) and less certain than we would choose. Further, the costs of action, while presumably far less than those that would result from inaction, are often more than sufficient to arouse constituencies in opposition, while the dangers lurking ahead lack the immediacy to easily animate support. The ultimate danger is that by remaining reliant on "the catastrophe theory of planning" in an era producing catastrophes of a magnitude greater than in the past, we can place our institutions in situations where precipitate action is the sole option—and it is then that our institutions themselves can be imperiled and individual rights overrun.

It is in that broader context that I recommend your careful reading and consideration of the report that follows. I think you will find that the matter of sea level rise is *not* an issue of the sort that Anwar Sadat had in mind when he jocularly said, "These are questions for the future generation." Just as the nations of the world are inexorably becoming more interdependent, so are the fates of the present and future generations. The issues raised in this report and the implications I have suggested are so important that we must begin to consider them today.

WILLIAM D. RUCKELSHAUS
Administrator, U.S. Environmental Protection Agency

Preface

Increasing concentrations of atmospheric carbon dioxide and other gases are expected to cause a global warming that could raise the sea several feet in the next century. In the spring of 1982 the U.S. Environmental Protection Agency organized a project to estimate the magnitude of future sea level rise, its effects, and the value of policies that prepare for these consequences. This book builds upon that effort.

Chapter 1 provides an overview of this book and is written for a general audience. Chapters 2 through 9 are for the more specialized reader. Chapter 10 provides the reactions of six prominent coastal decisionmakers to the rest of this book; it is written for the general reader. Chapter 1 spells out all measurements. Chapters 2 through 10, however, use abbreviations. For the convenience of the general reader, a list of abbreviations follows.

mm	milimeter	cm	centimeter
m	meter	m^2	square meter
m^3	cubic meter	km	kilometer
°C	degrees Celsius	in	inch
ft	foot	yd	yard
mi	mile	°F	degrees Farenheit

ACKNOWLEDGMENTS

It is a pleasure for us to thank the many persons who helped make possible this volume and the original work on which it is based.

The authors of the chapters in the book willingly and creatively participated in a project that involved a degree of interdisciplinary interaction new to all of us. They also cheerfully put their works into publishable form, often through many drafts.

Steven Seidel provided extensive organizational contributions to Chapters 1 and 8. Joan O'Callahan helped edit drafts of several of the chapters. Alex Wolfe provided information and contacts on the issue of hazardous waste disposal facilities in flood plains. He also reviewed drafts of Chapter

9. Peter Wilcoxen and Anthony Creamer assisted in identifying groups with a stake in sea level rise. Dr. Per Bruun reviewed early versions of chapters 1, 4, and 5.

Wynne Cougill assisted in editing and preparing the entire manuscript for submission. The quality and dedication of her work relieved the editors of an enormous burden for which mere gratitude seems inadequate. Charles Martin, now pursuing graduate work at Yale University, performed much of the programming of the models used in Chapters 7 and 8. Murray Kenney and Timothy Flynn performed much of the site research on Charleston and Galveston. Dennise Small supervised with great care the details of the Sea Level Rise Conference, held in Washington, D.C., in March 1983. Research assistants Vince Estrada, Lisa Maller, and Mark Wagner provided much and varied help. In the early stages of the project, Michael Goldman, Donald Greenberg, and Robin Sandenburgh helped shape its directions.

Local planners and officials from the communities in the Charleston and Galveston areas, too numerous to name, provided reports, insights, and data essential to the study. Two such persons deserve special thanks, for the economic analysis could not have been performed without the data they provided us. Robert W. Ragin, county assessor, county of Charleston, provided us with computer tapes of tax assessments and with essential explanation. Dr. Carlton Ruch of Texas A&M University provided us with data on property values in the Galveston study area, based on his important hurricane evacuation studies. Dr. Ruch also provided Stephen Leatherman with results of the new SLOSH computer hurricane model runs for the latter's use in Chapter 5.

At Research Planning Institute, Bruce N. Shibles assisted with the analysis. Jerry Cole, Cindy Heaton, and Starnell Percz assisted with graphics. At the University of South Carolina, Michael Todd assisted in mapping the Charleston area. At the University of Maryland, Beach Clow provided computer mapping of the Galveston area. Sue Gibbon and Barbara Young assisted with graphics.

We thank Margo Brown, who typed most of this book, for her cheerful work through countless revisions; she has our unbridled appreciation and respect. In addition, we express our appreciation to Andrea Calarco of the Goddard Institute of Space Studies, Zynda Rader and Diana Sangster of RPI, Pat Ledhan of the University of Maryland, and Kathy Miller of Lehigh University for typing various drafts of some chapters.

Michael Gibbs, a senior associate at ICF Incorporated, provided insight and input to the project far beyond his work reported in Chapter 7. He was a constant source of sound analysis and advice.

Finally, two individuals made invaluable contributions to the project. John Hoffman, who directs EPA's Strategic Studies Program, initially conceived of the project, obtained the resources for it, and helped guide it from start to finish. Joseph Cannon, then EPA's associate administrator for policy and resource management and now assistant administrator for air and radiation, provided essential moral and material support. Without their efforts, this book would not have been possible.

MICHAEL C. BARTH
JAMES G. TITUS

GREENHOUSE EFFECT AND SEA LEVEL RISE

Chapter 1

An Overview of the Causes and Effects of Sea Level Rise

James G. Titus and Michael C. Barth

with contributions by

Michael J. Gibbs, John S. Hoffman,
and Murray Kenney

INTRODUCTION

disscussion

The average person's view that sea level is constant is not shared by everyone, and for good reason. Petroleum companies and their geologists find oil on dry land once covered by prehistoric seas, and paleontologists find marine fossils on desert plains. Nevertheless, within the period of time relevant to most decisions, the assumption that sea level is stable has been appropriate. Only in a few cases have local changes in relative sea level due to land subsidence and emergence been large enough to have important impacts.

Recently, however, the view that current sea level changes are unimportant has been called into question. Coastal geologists are now suggesting that the thirty centimeter (one foot) rise in sea level that has taken place along much of the U.S. coast in the last century could be responsible for the serious erosion problems confronting many coastal communities.[1] Furthermore, according to the National Academy of Sciences, the expected doubling of atmospheric carbon dioxide and other greenhouse gases could raise the earth's average surface temperature 1.5-4.5°C (3-8°F) in the next century. Glaciologists have suggested that the sea could rise five to seven meters (approximately twenty feet) over the next several centuries from the resulting disintegration of the West Antarctic ice sheet.

A more immediate concern is that the projected global warming could raise the sea as much as one meter in the next century by heating ocean water, which would then expand, and by causing mountain glaciers and parts of ice sheets in West Antarctica, East Antarctica, and Greenland to melt or slide into the oceans. Thus, the sea could reach heights unprecedented in the history of civilization. Until this effort, no one had attempted to forecast sea level rise in specific years or determine its importance to today's activities.*

A rise in sea level of even one meter during the next century could influence the outcomes of many decisions now being made. In the United States, thousands of square miles of land could be lost, particularly in low-lying areas such as the Mississippi Delta, where the land is also subsiding at approximately one meter per century. Storm damage, already estimated at over three billion dollars per year nationwide, could also increase, particularly along the well-developed and low-lying Atlantic coast. Finally, a rising sea will increase the salinity of marshes, estuaries, and aquifers, disrupting marine life and possibly threatening some drinking water supplies. Fortunately, the most adverse effects can be avoided if timely actions are taken in anticipation of sea level rise.

Although action may be taken to limit the eventual global warming from rising atmospheric CO_2, the warming expected in the next sixty years and the resulting rise in sea level are not likely to be prevented. Most CO_2 emissions are released by burning fossil fuels. Because these fuels are abundant and relatively inexpensive to produce, a voluntary shift to alternative energy sources is very unlikely. Regulatory action that would effectively limit CO_2 concentrations is also unlikely. Such actions by any one nation, even the United States, could delay the effects of increasing concentrations of CO_2 by a few years at most, while imposing competitive disadvantages on the nation's industries. Emissions of other trace gases (such as chlorofluorocarbons and methane) could add significantly to the projected global warming. Furthermore, the uncertainties surrounding the impacts on climate currently make it impossible to determine whether preventing the global warming would provide a net benefit to the world or to individual nations. Finally, even if emissions are curtailed, global temperatures and sea level will continue to rise for a few decades as the world's oceans and ice cover come into equilibrium.

Although preventing a global warming would require a worldwide

---

*Editors' note: After the submission of this manuscript, the NAS released a projection that sea level could rise seventy centimeters by 2080, not including the impact of Antarctica (see Revelle, 1983).

consensus, responding to its consequences would not. Communities can construct barriers or issue zoning regulations; companies and individuals can build on higher ground; and environmental agencies can take measures to reserve dry lands for eventual use as biologically productive wetlands.

To meet the challenge of a global warming, society will need accurate information concerning the likely effects of sea level rise. Unfortunately, communities, corporations, and individuals do not by themselves have sufficient resources or incentives to undertake the basic scientific research required to reduce existing uncertainties. This responsibility falls upon national governments throughout the world. Only their efforts can provide the information that decision makers will need.

This book is based on interdisciplinary efforts that the United States Environmental Protection Agency (EPA) initiated to encourage the development of information necessary to adapt to sea level rise. In the spring of 1982, EPA organized a project aimed at developing methods to study the effects of sea level rise and estimate the value of policies that prepare for this rise. The project proceeded in the following steps, as illustrated in Figure 1-1.

Available scientific research was used to project conservative, low, medium, and high scenarios of global sea level rise through 2100.

The scenarios were adjusted for local trends in subsidence to yield local sea level rise scenarios through 2075 for two case study sites—Galveston, Texas, and Charleston, South Carolina.

Economic and environmental scenarios were developed for the two case study sites, assuming no rise in sea level.

The physical effects of sea level rise for the case study areas were estimated.

The economic effects of sea level rise if it were not anticipated were estimated.

Options for preventing, mitigating, and responding to the effects of sea level rise were developed.

The economic effects of sea level rise if it were anticipated were estimated.

The value of anticipatory actions and better projections of sea level rise was assessed.

Given the broad range of disciplines encompassed in this effort and the range of individuals to whom it might be of interest, this introductory chapter provides an overview of the entire project, written for the general reader. Chapters 2 through 10 explore the issues in more detail.

Chapter 2 summarizes the scientific evidence on the relationship between rising CO_2 concentrations and global temperatures.

Chapter 3 sets forth the range of estimates for sea level rise that underlie the remainder of the analysis.

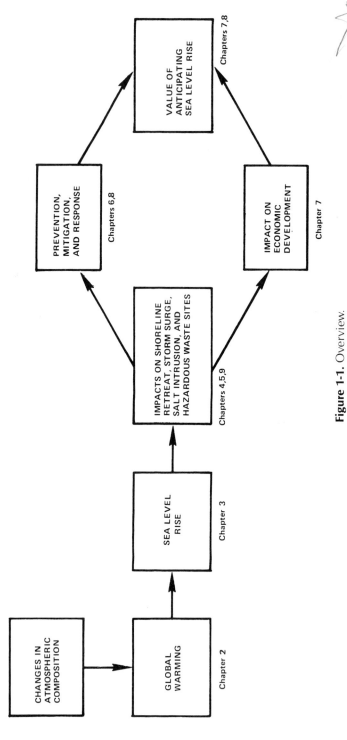

Figure 1-1. Overview.

CHANGES IN ATMOSPHERIC COMPOSITION

GLOBAL WARMING

Chapter 2

SEA LEVEL RISE

Chapter 3

IMPACTS ON SHORELINE RETREAT, STORM SURGE, SALT INTRUSION, AND HAZARDOUS WASTE SITES

Chapters 4,5,9

PREVENTION, MITIGATION, AND RESPONSE

Chapters 6,8

IMPACT ON ECONOMIC DEVELOPMENT

Chapter 7

VALUE OF ANTICIPATING SEA LEVEL RISE

Chapters 7,8

Chapter 4 presents the method and results of an analysis of the effects of sea level rise on the Charleston area. The chapter projects the two causes of shoreline retreat, inundation and erosion, as well as changes in flood levels and salt intrusion into aquifers.

Chapter 5 presents an analysis similar to that in Chapter 4, using somewhat different methods for the Galveston Bay area.

Chapter 6 catalogues the potential engineering responses to sea level rise, their costs, and their potential effectiveness.

Chapter 7 presents the methods, data, and results of an economic impact analysis of the physical effects of sea level rise at the two case study sites as well as an analysis of the benefits of anticipating the rise in terms of reducing adverse impacts.

Chapter 8 examines policy options for resort communities adapting to sea level rise and the decisions that property owners on Sullivans Island, South Carolina, would face after a major storm.

Chapter 9 indicates how sea level rise may affect existing hazardous waste facilities and implications for the regulation of proposed facilities.

Chapter 10 presents the reactions of six potential users of this information delivered to a conference on sea level rise in Washington, D.C., on March 30, 1983. In the first comment, Dr. Sherwood Gagliano discusses Chapters 4, 5, and 6, as well as his experience with relative sea level rise in Louisiana. The other comments present a broad range of views on the technical and social implications of sea level rise.

Progress in understanding sea level rise and the most appropriate ways to respond will require discussions within and between diverse disciplines including biology, climatology, economics, engineering, geology, geography, hydrology, meteorology, and urban planning. The most important needs are: less uncertainty in the range of sea level rise estimates; better methods to assess the physical effects of sea level rise; better methods to estimate economic impacts on specific communities and private-sector firms; assessments of the actions that could be taken in response to, and in anticipation of, sea level rise; greater awareness on the part of potentially affected parties; and better estimates of the potential savings from anticipating sea level rise.

We have only begun to determine the degree to which research should be accelerated to produce better forecasts of sea level rise. Such an assessment is necessary to ensure that government efforts to address sea level rise are allocated a level of resources commensurate with the potential benefits of such efforts. The case studies reported here indicate that Charleston and Galveston could save hundreds of millions of dollars by preparing for sea level rise. If additional analyses are consistent with the findings of the case studies, then the value of better forecasts would easily justify the substantial costs of developing them. More research should be undertaken to confirm our findings; because of the time it will take to improve sea level rise estimates, an evaluation of the appropriate priority for such research should not be delayed.

This book provides a framework for understanding the importance of sea level rise. The methods developed and applied to Galveston and Charleston can be used for other jurisdictions. They can also be used by corporations, municipalities, or states to evaluate individual project decisions in the coastal zone. Parties that could be affected by sea level rise should determine whether the impacts will require changes in their operations and the importance of better forecasts.

We hope that this book proves to be more than a collection of useful scientific papers. We believe that it raises important policy issues that warrant the attention of all citizens, not just those who allocate research budgets, issue government regulations, and make investment and locational decisions. Responding to the challenge of a rising sea will require better assessments and public awareness of the future rate of sea level rise, the likely effects, and options for slowing the rise or adapting to it. Our goal is to accelerate the process by which these issues are resolved.

SEA LEVEL, CLIMATE, AND CARBON DIOXIDE EMISSIONS

The rise and fall of sea level is influenced by both geological and climatic factors. Changes in mid-ocean ridge systems may have been responsible for a drop in sea level of three hundred meters (one thousand feet) over the last eighty million years (Hays and Pitman, 1973).[2] Even today, emergence and subsidence of land can have a noticeable effect on local sea level. For example, Louisiana is currently losing over one hundred square kilometers (approximately fifty square miles) of land per year, largely because of subsidence estimated at one meter per century (Boesch, 1982). In contrast, emergence has caused difficulty for Finnish port authorities facing progressively shallower harbors.

Geological events affecting sea level are, however, generally slow and unlikely to accelerate. Although this has generally been true for climatic changes in the past, the future may be different. This section looks at the relationship between sea level and climate, explaining how rising atmospheric concentrations of carbon dioxide can raise the earth's average surface temperature and thereby dramatically change both climate and sea level.

The Relationship between Climate and Sea Level

Climate influences sea level in two ways: by moving the earth's water between glaciers resting on land and the oceans and by changing the temperature of the ocean water and thus its volume. If all the glaciers in Antarctica and Greenland melted, sea level would rise more than seventy meters (over two hundred feet). In the past, enough ocean water has accumulated in glaciers to lower sea level about one hundred and fifty meters (five hundred feet).

Although complete melting of land-based glaciers would take thousands of years, partial melting could raise sea level as much as a meter in the next century. Furthermore, glaciers grounded under water could disintegrate more quickly. Two leading glaciologists have estimated that the entire West Antarctic ice sheet (the largest marine-based glacier) could enter the oceans in two hundred years (Hughes et al., 1979) and five hundred years (Bentley, 1980) raising sea level five to seven meters (about twenty feet). Although a complete disintegration of this marine-based glacier will not occur in the near future, parts of it and other ice-fields, as well as mountain glaciers, could be vulnerable in the next century.

Because water expands when heated, a warmer climate could raise the sea even without any contribution from glaciers. Although a warming of the entire ocean would take several centuries, the upper layers could warm and raise sea level as much as a meter by 2100. This shorter-term effect of a global warming is frequently overlooked.

Past Trends in Climate and Sea Level

For the last two million years and probably longer, sea level and climate have fluctuated together in cycles of 100,000 years. These cycles are caused by changes in solar irradiance due to cyclic changes in the tilt of the earth's axis. During ice ages, the earth's average temperature has been about 5°C (9°F) colder than at present, with glaciers covering major portions of the continents. During the Last Glacial (12,000-20,000 years ago), sea level was approximately one hundred meters (over three hundred feet) lower than today. During previous ice ages it may have been one hundred and fifty meters lower (Donn et al., 1962).

During the warm interglacial periods, temperatures and the sea have risen to approximately the levels of today. There is no evidence that the land-based glaciers in Greenland and Antarctica have ever completely melted in the last two million years. However, glaciologist J. H. Mercer (1972) has suggested that the West Antarctic ice sheet has completely disappeared, with sea level rising five to seven meters above its present level, probably during the last interglacial period (115,000 years ago). From the end of the Last Glacial until about six thousand years ago, sea level rose approximately one meter per century.

In the last century, tidal gauges have been available to measure sea level at specific locations. Studies combining these measurements to determine trends in worldwide sea level have concluded that it has risen ten to fifteen centimeters (four to six inches) in the past century (Fairbridge and Krebs, 1962; Gutenberg, 1941; Lisitzin, 1974; Barnett, 1983; Gornitz et al., 1982). At least part of this rise can be explained by the warming trend of 0.4°C in the last century and the resulting thermal expansion of the upper layers of the ocean (Gornitz et al., 1982). The remainder may be due to a small amount of glacial melting and a delayed response of deep-ocean waters to longer-term warming trends.

The Greenhouse Effect and the Prospect of Global Warming

Although climate and sea level have been relatively stable in recent centuries, the next century may be very different. In the past, the delicate balance of the global climatic system has evolved slowly as its various determinants shifted. Current activities, however, are altering this balance.

Man is reversing millions of years of natural evolution by putting back into the atmosphere carbon that had been sequestered over the ages as fossil fuels. Atmospheric concentrations of CO_2 are likely to double, and possibly triple, by 2100. Because no historical precedent exists, reasonable expectations about future climate must be based on scientific evidence, not geological records. After evaluating the available evidence, the National Academy of Sciences concluded that a doubling of atmospheric concentrations of CO_2 would warm the earth's average surface temperature 1.5-4.5°C (2.7-8.1°F) (Charney, 1979; Smagorinsky, 1983).

The greenhouse effect of the atmosphere has never been doubted. Most of the sun's radiation is visible light, which passes through the atmosphere largely undeterred. When the radiation strikes the earth, it warms the surface, which then radiates the heat as infrared radiation. However, atmospheric CO_2, water vapor, and some other gases absorb the infrared radiation rather than allow it to pass undeterred through the atmosphere to space. Because the atmosphere traps the heat and warms the earth in a manner somewhat analogous to the glass panels of a greenhouse, this phenomenon is generally known as the "greenhouse effect."[3] Without this effect, the earth would be 33°C (60°F) colder than it is currently.

The extent to which CO_2 absorbs heat has been known for almost a century (Arrhenius, 1896). In Chapter 2, Hansen et al. show that a doubling of atmospheric CO_2 would raise the average temperature 1.2°C (2.0°F) if nothing else in the earth's climatic system changed. However, many parts of the climate will change, amplifying the direct impact of CO_2. Because these changes are not completely understood, the total warming is difficult to estimate. The current uncertainty surrounding the impact of CO_2 on average temperature is centered around these climatic "feedbacks," not the direct warming from CO_2. Evidence of some of these feedbacks is so strong that the National Academy of Sciences has concluded that the warming will be at least 1.5°C.

The most important feedback will result from the warmer atmosphere's ability to retain more moisture. Because water vapor also absorbs infrared radiation, additional heating will result. Hansen et al. estimate that doubled CO_2 would increase the atmosphere's water vapor content 30 percent, heating the earth an additional 1.4°C.

Another important positive feedback concerns the impact of snow and ice cover on the earth's albedo, the extent to which it reflects sunlight. Ice and snow reflect most of the sun's radiation, while water and soil absorb it. An increase in surface temperatures would melt snow cover on land and floating ice and thereby allow the earth to absorb energy that would otherwise be reflected back into space. Hansen et al. estimate an additional warming of 0.4°C from the albedo effect.

A feedback that is less understood is the impact of a global warming on clouds, which also reflect sunlight into space. The effects of clouds on the earth's albedo depend on their heights and other properties, as well as the extent of cloud cover. Thus, the impact of a global warming on clouds is somewhat uncertain. Nevertheless, with somewhat less confidence, Hansen et al. estimate a 2 percent reduction in cloud cover and a resulting warming of 0.5°C. They also estimate that increases in cloud height would result in an additional warming of 0.5°C, for a total impact of 1.0°C from clouds.

Although the increase in the average temperature of the earth is a convenient shorthand description of CO_2-induced climatic change, it masks important regional implications. Most researchers agree that polar temperatures would increase two to three times the earth's average increase. The world's climate depends largely on circulation patterns by which the atmosphere and the oceans transport heat from warm to cold regions. As a result, any significant change in the difference between equatorial and polar temperatures could dramatically affect climatic patterns. A particularly important effect of these changes will be shifts in annual and seasonal precipitation and evaporation, with some areas gaining and others losing. Furthermore, because hurricanes require an ocean temperature of 27°C (80°F) or warmer, a global warming could allow hurricanes to form at higher latitudes and during a greater part of the year. Although these changes could be important to coastal communities, they have not been examined in this study.[4]

Increasing Atmospheric Concentrations of Carbon Dioxide and Other Greenhouse Gases

Although the climatic change that would result from CO_2 emissions is poorly understood, there is complete agreement that CO_2 concentrations are increasing. The measured concentration of CO_2 in the atmosphere increased from 315 parts per million in 1958 to 339 ppm by 1980 (Keeling, 1982). Estimates from tree rings suggest that the concentration was approximately 280 ppm in 1860.

Approximately one-half the CO_2 released by combustion of fossil fuels has remained in the atmosphere. It is generally believed that most of the remaining CO_2 has dissolved into the oceans. Although tropical deforestation and cement production also result in CO_2 emissions, their contributions have been and will continue to be much less important.

In the next few decades, CO_2 emissions are unlikely to be curtailed, either voluntarily or by regulation. The world's infrastructure is built around fossil fuels. The cost of using coal, gas, and oil is low compared with nuclear and solar power, and this relative cost advantage is expected to continue. Therefore, a voluntary reduction in CO_2 emissions is unlikely.

The only governmental action that could successfully reduce CO_2 emissions would be to curtail the use of fossil fuels. Emission controls (scrubbers) for CO_2 from power plants would at least quadruple the cost of electricity (Albanese, 1980). For smaller users of fossil fuels, such as homes and motor vehicles, control is not even feasible. Other plans, such as sequestering carbon in massive tree plantings, are even less plausible (Greenberg, 1982).

Even if political leaders decide to take drastic actions to limit worldwide consumption of fossil fuels, it is probably already too late to prevent significant rises in global temperatures and sea level. A recent study by EPA investigated the impact of drastic energy policy changes on the expected timing of a greenhouse warming (Seidel and Keyes, 1983). The authors concluded that such policies could have important impacts by 2100, but would not substantially delay the 2°C warming expected by 2040. They estimated that a 300 percent tax on fossil fuels would delay the 2°C warming by only five years, and that even a worldwide ban on coal, shale oil, and synthetic fuels would delay the warming by only twenty-five years, if implemented by 2000. Furthermore, such a ban would delay the rise in sea level expected through 2040 by only twelve years.[5]

The political feasibility of instituting such a ban by 2000 is also doubtful, because only a worldwide agreement to curtail emissions could be successful. Any individual nation that curtails its own emissions will delay the day when CO_2 concentrations double by a few years at most. (This delay would be even less if the resulting drop in energy prices induced other nations to increase their own consumption.)[6] Furthermore, because energy costs would increase for any nation that curtailed its emissions, that nation's industries would be placed at a competitive disadvantage compared with those of the rest of the world. The failure of most other nations to follow the United States' lead in banning chlorofluorocarbons in spray cans, where the costs were very minor, indicates that reaching a worldwide consensus on curtailing emissions is extremely difficult. Finally, political leaders would require proof that such a policy would be more beneficial than adapting to higher CO_2 levels. Such proof will probably remain impossible to provide for the forseeable future.

Several other gases emitted by human activities also absorb infrared radiation, and would thus contribute to a global warming. The most significant of these trace gases are methane, nitrous oxide, and chlorofluorocarbons. As Hansen et al. discuss in Chapter 2, emissions of these gases added 50-100 percent to the greenhouse effect from CO_2. Although less is known about the future importance of these gases, emissions of some of them, particularly chlorofluorocarbons, may grow much faster than CO_2 (Palmer et al., 1980).

The impact of increasing concentrations of greenhouse gases will almost certainly be an unprecedented global warming. Some people have suggested that this warming may be offset because the earth would otherwise be entering a cool period. However, a natural cooling would take place over tens of thousands of years and is thus unlikely to significantly offset the global warming in the next century. Even a drastic increase in volcanic activity would offset less than 10 percent of the projected rise in sea level (Hoffman et al., 1983).

Estimating the Magnitude of the Greenhouse Effect

In the last few decades, mathematical models have been developed to estimate the impact of CO_2 on climate. Two of the most complete climate models, those of Hansen et al. and Manabe and Stouffer (1980), estimate the warming from doubled CO_2 to be 4°C and 2°C, respectively.[7] In Chapter 2, Hansen et al. discuss the differences between these models, and other evidence supporting estimates of the magnitude of the greenhouse effect. They conclude by estimating that in the 1990s the warming from the greenhouse effect will exceed the fluctuations that have occurred in this century, laying to rest any remaining doubts about the importance of the greenhouse effect.

One of the most important differences between the two models is that Manabe and Stouffer assume that the behavior of clouds would not change, while the Hansen et al. model predicts it to be an important positive feedback. The former model also assumes that less sea ice exists and therefore that the albedo effect will be less significant. Finally, the Manabe and Stouffer model assumes that the atmosphere transports all heat from equatorial to polar regions, while Hansen et al. assume that ocean currents also transport heat. These differences cause Manabe and Stouffer's estimate to be lower than that of Hansen et al.

Hansen et al. identify three types of evidence that support estimates of the magnitude of the greenhouse effect: temperatures on other planets, recent global temperature trends, and long-term climate cycles. Compared with the earth, Mars has lower concentrations and Venus higher concentrations of CO_2 and other greenhouse gases. Hansen et al. show that temperature differences between these planets are well-explained by the greenhouse effect, not merely by their distances from the sun. For example, without the greenhouse effect, Venus would be approximately the same temperature as the earth. However, because the planet's atmosphere is mostly CO_2 and traps infrared radiation more than one hundred times as efficiently as the earth's atmosphere, Venus is 400°C hotter.

Hansen et al. show that their model's predictions are also consistent with historical evidence. In the past century, global temperatures have increased 0.4°C, with 0.1°C fluctuations from decade to decade. Hansen et al. show that much of the variation in temperature can be explained by their model when the impacts of CO_2 and volcanoes are considered. Another type of historical evidence is the ability of the models to explain climatic periods from long ago. Over the last 18,000 years, the earth's average temperature has increased 4°C as the ice covering much of North America, Europe, and Asia retreated. Hansen et al. show that the changes in ice cover used by their model to predict the warming from CO_2 is consistent with the changes in ice cover that have occurred in the last 18,000 years.

SEA LEVEL RISE SCENARIOS

Faced with the consequences of a global warming, coastal decision makers would like to have a precise projection of sea level rise. Unfortunately, because of the large degree of uncertainty in many of the factors influencing sea level, available scientific knowledge is inadequate to generate a precise forecast.

Nevertheless, in Chapter 3, Hoffman argues that available knowledge is sufficient to estimate the likely range of sea level rise in the next century. For each of the factors influencing sea level rise, he consulted the experts and the literature to determine conservative and high estimates. He then linked various combinations of these estimates to produce scenarios of worldwide sea level rise ranging from conservative to high.

Scenario Building

Figure 1-2 illustrates the relationships among the factors influencing sea level rise that Hoffman considered. Several different models representing these components were used to generate over 90 scenarios. From these, a conservative, a mid-range low, a mid-range high, and a high scenario were identified.

The major factors influencing sea level that Hoffman considered were: CO_2 emissions; fraction airborne (the fraction of CO_2 emissions that remains in the atmosphere); concentrations of other trace gases; climate sensitivity (global warming resulting from increases in atmospheric concentrations of CO_2 and trace gases); thermal expansion of ocean water; and snow and ice contributions.

For the first five factors, Hoffman specified a conservative, a mid-range, and a high assumption. For snow and ice contributions, he used only a high and a low assumption.

Carbon Dioxide Emissions.   The World Energy Model of the Institute for Energy Analysis was run under a variety of assumptions regarding population growth, economic activity, and the relative costs of various sources of energy to produce scenarios of CO_2 emissions (Institute for Energy Analysis, 1982). Based on the work of Keyfitz et al. (1983), all scenarios assumed that world population achieved zero growth by 2075. The high scenario assumed that per capita economic growth decreased from 3.5 percent per year in 1980 to 2.2 percent by 2100. These rates are lower than experienced by the world economy in the last thirty years. For the conservative scenario, growth will diminish from 2.2 percent in 1980 to 1.7 percent in 2100. All scenarios assumed that energy efficiency improves, and the conservative scenario also assumed that the cost of producing nuclear power was reduced 50 percent in 1980. As a result of these assumptions, CO_2 emissions would grow at average rates of 1.7 percent, 2.0 percent, and 2.3 percent per year from 1980 to 2100, for the conservative, mid-range, and high scenarios, respectively.

Fraction Airborne.   Two methods were used to determine the percentage of carbon emissions that remain in the atmosphere (i.e., the fraction airborne). In the conservative scenario, the historical average of 53 percent was used. In the mid-range and high scenarios, the Carbon Cycle

Model of the Oak Ridge National Laboratories (ORNL) was used (ORNL, undated). This model simulates the movement of carbon among the biosphere, oceans, and atmosphere, taking into account decay, oxidation, and other biochemical actions. Largely because the upper layers of the

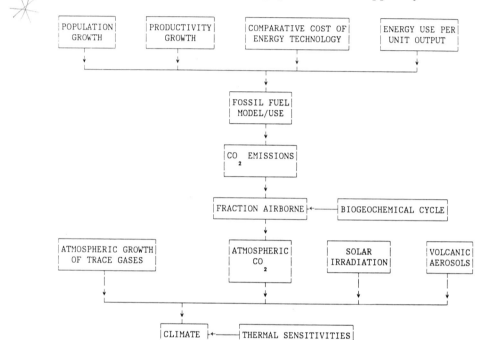

Figure 1-2. Basis for scenarios. For each factor or relationship, high and low assumptions were developed using the published literature.

ocean would approach saturation as warmer surface temperatures reduced vertical mixing of the oceans, the model predicted that the rate of atmospheric retention of CO_2 would grow from 60-80 percent by 2100. As a result, atmospheric concentrations of CO_2 would double by 2055 in the high scenario and by 2085 in the conservative scenario.

Concentrations of Other Greenhouse Gases. Knowledge of the origins and fates of other gases that absorb infrared radiation is insufficient to project their atmospheric concentrations in the same manner as was done for CO_2. Hoffman only considered four of the most important trace gases: methane, nitrous oxide, and two chlorofluorocarbons, $CFCl_3$ (R-11) and CF_2Cl_2 (R-12).

Because of the ozone depletion potential from chlorofluorocarbons, all scenarios assumed that emissions would not increase after 2020. From 1980 until that date, the conservative, mid-range, and high scenarios assumed that emissions of these gases would increase linearly by 0.7, 2.5, and 3.8 percent of the 1980 level, respectively, each year. Concentrations of these gases were calculated by assuming the half-lives of $CFCl_3$ and CF_2Cl_2 to be 60 and 120 years, respectively. For nitrous oxide and methane, Hoffman projected atmospheric concentrations directly. The three scenarios assumed that methane concentrations would increase geometrically by 1.0, 1.5, and 2.0 percent per year, and that nitrous oxide concentrations would increase by 0.2, 0.45, and 0.7 percent per year.

Climate Sensitivity. Hoffman used the National Academy of Sciences' estimated range of the impact of a CO_2 doubling on average surface temperature (Charney, 1979; Smagorinsky, 1982). The conservative scenario assumed that the average surface temperature would increase 1.5°C (2.7°F), the mid-range scenarios assumed 3.0°C (5.4°F), and the high scenario used 4.5°C (8.1°F). Given these assumptions about the impact of a doubling, Hoffman projected year-by-year increases in temperature using the increases in greenhouse gases with an equation fit to the results of a climatic model that includes heat transfer from the atmosphere to the oceans.[8]

Thermal Expansion of Oceans. Although surface waters would warm quickly with rising global air temperatures, the downward transport of heat into deeper ocean layers would warm them much more slowly. Hoffman employed an ocean model that uses diffusion as a surrogate for all heat transport processes to estimate heat transported into the top one thousand meters (3,200 feet) of the ocean. All scenarios used coefficients based on interpretation of data on ocean tracers.[9] Because glacial melting and climate change might alter ocean currents drastically, special-case scenarios were also developed.[10] These extreme assumptions did not have large effects on the resulting projections of sea level rise. The expansion of water was computed at each depth using standard coefficients of expansion for the temperature, pressure, and salinity of a globally averaged column of water.

Snow and Ice Contributions. Very little work has been done concerning the impact of a global warming on deglaciation. As a result, Hoffman acknowledges that his assumptions about the impact of global warming on glaciers constitute the weakest part of the analysis. He notes that in the last century, a global warming of 0.4°C (0.7°F) would be sufficient to explain a 5 cm (2 in) rise in sea level from thermal expansion. However, various authors have estimated that the actual rise was 10-15 cm (4-6 in). Therefore, factors other than thermal expansion—most likely snow and ice contributions from land—accounted for the other 5-10 cm rise. Given the absence of glacial process models, Hoffman assumed that this relationship would persist.

Therefore, the conservative and mid-range low scenarios assumed that the rise in sea level from deglaciation would equal the contribution from thermal expansion, while the mid-range high and high scenarios assumed that it would be twice the contribution. Hoffman notes that these assumptions were consistent with estimates of melting derived from the Hansen et al. three-dimensional global climate model, which simulates world climate on an hour-by-hour basis in a manner very close to observed conditions. Nevertheless, he argues that this aspect of the scenarios should be improved as soon as possible by using glacial process models.*

Results of Sea Level Rise Scenarios

Table 1-1 illustrates Hoffman's results for global sea level rise through 2100. Under the high scenario, sea level will rise about 17 cm (6.7 in) by 2000, 117 cm (3.8 ft) by 2050, and 345 cm (11.3 ft) by 2100. Under the conservative scenario, sea level will rise 4.8 cm (2 in) by 2000, 24 cm (9.4 in) by 2050, and 56 cm (22 in) by 2100. Because of local subsidence, most of the Atlantic and Gulf coasts of the United States can expect the sea to rise 15-20 cm more in the next century than these figures indicate.

Tables 1-2 and 1-3 show the sea level rise scenarios that Kana et al. and Leatherman used in the Charleston and Galveston case studies. These scenarios differ from Hoffman's scenarios for several reasons. First, Hoffman made several improvements in his scenarios after Kana et al.

---

*Editors' Note: Subsequent analysis has led Hoffman to conclude that the mid-range low scenario is more likely than the mid-range high scenario.

Table 1-1. Worldwide Sea Level Rise Scenarios, 1980–2100 (in cm and ft above 1980 levels)

Scenario	2000	2025	2050	2075	2100
Conservative	4.8 (0.16)	13.0 (0.43)	23.8 (0.78)	38.0 (1.2)	56.2 (1.8)
Mid-range Low	8.8 (0.29)	26.2 (0.86)	52.3 (1.7)	91.2 (3.0)	144.4 (4.7)
Mid-range High	13.2 (0.43)	39.3 (1.3)	78.6 (2.6)	136.8 (4.5)	216.6 (7.1)
High	17.1 (0.56)	54.9 (1.8)	116.7 (3.8)	212.7 (7.0)	345.0 (11.3)

Table 1-2. Sea Level Rise Scenarios for Charleston, 1980–2075 (in cm, with ft in parentheses)

		Year	
Scenario[a]	1980	2025	2075
Baseline	0	11.2 (0.4)	23.8 (0.8)
Low	0	28.2 (0.9)	87.6 (2.9)
Medium	0	46 (1.5)	159.2 (5.2)
High	0	63.8 (2.1)	231.6 (7.6)

Source: Global sea level rise scenarios are from Chapter 3, modified to reflect local conditions based on the historical trend for Charleston. (S. D. Hicks et al., 1983, *Sea Level Variations for the United States, 1855–1980,* technical report, Rockville, Md., NOAA, Tides and Water Levels Branch)

[a]Baseline scenarios for each year reflect present trends. Other scenarios reflect accelerated sea level rises at various rates.

Table 1-3. Sea Level Rise Scenarios for Galveston, 1980–2075 (in cm, with ft in parentheses)

		Year	
Scenario	1980	2025	2075
Baseline	0	13.7 (0.45)	30.0 (0.98)
Low	0	30.7 (1.0)	92.4 (3.0)
Medium	0	48.4 (1.6)	164.5 (5.4)
High	0	66.2 (2.2)	236.9 (7.8)

Source: See Chapter 5.

and Leatherman had completed their chapters.[11] Second, Kana et al. and Leatherman adjusted the global scenarios to account for local conditions. Third, instead of using the conservative scenario, the case studies used a baseline scenario calculated by extrapolating past trends of global sea level rise and using judgment regarding local trends. Finally, the "mid-range low" scenario is called "low" and the "mid-range high" scenario was replaced by a "medium" scenario equal to the average of the high and low scenarios.

Research Necessary to Reduce Uncertainty

Opportunities are available for substantially reducing the major uncertainties regarding sea level rise. However, better knowledge will require considerably greater research expenditures than are currently being made. Moreover, a mission-oriented management will be needed to induce the necessary coordination between researchers who normally work apart in such diverse fields as climatology, oceanography, glaciology, and biogeochemistry.

Of the major factors considered in Hoffman's scenarios, insufficient research is currently being undertaken on concentrations of trace gases, deglaciation, and incorporation of the oceans into climate models. Although there is a modest amount of ongoing research to determine the likely impact of CO_2 emissions on average global temperatures, these activities are not driven by a sense of urgency that is based on the need to produce year-by-year estimates that are useful for decision makers.

In researching trace gases, the short-term priority should be to identify all the sources and sinks, both current and future, that will influence concentrations of chlorofluorocarbons, methane, nitrous oxide and other important gases neglected in this study. Over the longer term, biogeochemical models that accurately represent the atmospheric, oceanic, and terrestrial processes that control the levels of these gases need to be developed.

Climate models that incorporate realistic geography, realistic heat uptake and transport by the oceans, and the feedback effects of melting, evaporation, sublimation, and snowfall in polar regions should be run on a year-by-year basis as soon as possible in order to provide estimates of the time path and geography of climate change. A major effort to include better ocean models should be the highest long-term priority. A better representation of polar processes in global climate models is also necessary.

The response of glaciers to a global warming is the least understood of the major factors that will determine sea level rise. In the short run, global climate modelers, southern ocean oceanographers, and glaciologists can produce scenarios of meltwater runoff and deglaciation that complement the scenarios of thermal expansion developed by Hoffman. In the longer term, a greater data collection will be needed. Without the better observations necessary to build and validate models, it will be impossible to provide reliable and precise estimates. In the next decade, more complete models of icefields should be developed, based on the specific topography of each field. Experiments such as towing icebergs into warmer water could also be undertaken to provide additional insights into the behavior of glaciers under radically different conditions. Observational programs using satellite data to track the advance and retreat of glaciers should also be undertaken. Together, these efforts can greatly improve the precision of estimates of snow and ice contributions to sea level rise. Finally, models of thermal expansion that consider longitude and latitude should replace the one-dimensional model Hoffman used. Over the longer run, models of ocean circulation capable of considering the impacts of global warming and deglaciation on ocean mixing, and thus heat uptake, should be developed.

The challenge of advancing our knowledge will require careful management of research. Only if sustained long-term support is given to interdisciplinary scientific teams can accelerated research speed the development of necessary information and narrow the range of plausible sea level rise scenarios. The sporadic stop-and-start efforts and the support of individuals or groups working in isolation that have characterized many recent efforts are not likely to be sufficient for this challenge.

EFFECTS OF SEA LEVEL RISE

This section describes the physical and environmental effects of sea level rise, and the activities that can be undertaken to prevent or adapt to these effects.

The Physical Consequences of Sea Level Rise

The physical consequences of sea level rise can be broadly classified into three categories: shoreline retreat, temporary flooding, and salt intrusion. The most obvious consequence of a rise in sea level would be permanent flooding (inundation) of low-lying areas. A sea level rise of a few meters would inundate major portions of Louisiana and Florida, as well as beach resorts along the coasts. Marshes and low-lying flood plains along rivers and bays would also be lost.

Many coastal areas with sufficient elevation to avoid inundation would be threatened by a different cause of shoreline retreat: erosion. In fact, the current trend of sea level rise may be causing the serious erosion that is taking place in many coastal resorts (New Jersey, 1981; Pilkey et al., 1981). The constant attack of waves causes beaches to take a particular profile, which fluctuates seasonally. Winter storms erode the upper beach and deposit sand offshore, and the calmer spring and summer waves redeposit the sand and restore the beach. However, a rise in sea level alters the relationship of the shore profile to the water level (see Figure 1-3). Because the water near the beach is deeper than before, more energy is required to move the offshore sand back to the beach. Consequently, some of the material deposited offshore by winter storm waves remains offshore, and a portion of the beach is lost (Bruun, 1962; Schwartz and Fisher, 1979).

Another cause of beach erosion from sea level rise is overwash and the resulting landward migration of coastal barriers (Massachusetts, 1981; U.S. Department of Interior, 1983). Many American beach resorts lie on narrow islands and spits (peninsulas with the ocean on one side and a bay on the other). Rather than erode them in place, overwash processes cause the islands and spits to migrate landward in a fashion similar to a tank tread. These processes take place during storms and raise islands as well as move them landward. Although this process may protect the barrier itself, property on the seaward side may be totally lost.

Increased storm damage is an economically important result of sea level rise. Wind and low pressure during hurricanes and other storms cause the water level in an area to rise temporarily, sometimes by several meters. The Federal Emergency Management Agency classifies these storms in terms of both their frequency and the magnitude of the elevation in water level—the latter phenomenon is known as storm surge. Regions with 1 percent and 10 percent chances of flooding in a given year are designated as 100-year and 10-year flood zones, respectively. Existing

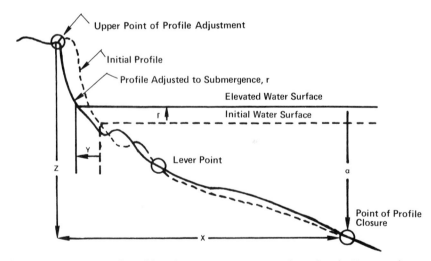

Figure 1-3. Concept of profile adjustment to increased sea level. Given a shore profile at equilibrium and a rise in water level, Bruun's Rule (1962) states that beach erosion occurs in order to provide sediments to the shore bottom so that the shore bottom can be elevated in proportion to the rise in water level (a). The volume of sediment eroded from the beach (V) is equal to the volume of sediment deposited on the shore bottom (V'). *(Source: after E. B. Hands, 1981. Predicting Adjustments in Shore and Offshore Sand Profiles on the Great Lakes. CERC Technical aid 81-4, Ft. Belvoir, Va.: Coastal Engineering Research Center.)*

development is often predicated on the basis of these flood frequencies. An increase in sea level would increase the water level during a flood by approximately the amount of sea level rise, bringing new areas into the flood zones. Higher water tables also exacerbate flooding by decreasing the ability of land to drain stormwaters. Finally, erosion and deeper water could subject new areas to damaging storm waves.

Sea level rise also causes the salt content of aquifers and estuaries to migrate landward. In coastal aquifers, a layer of freshwater floats on top of the heavier saltwater. The saltwater generally forms a wedge such that the farther inland (the higher the water table), the farther below ground is the boundary between fresh and saltwater. Where sea level rise results in a landward movement of the shoreline, this boundary will move inland as well. Because the level of the water table is itself determined by sea level, a rise in sea level causes the freshwater/saltwater boundary to rise. The landward and upward shift of this boundary implies that certain freshwater wells may become salty. Overpumping of coastal aquifers also has resulted in salt intrusion, however, and in many instances this problem dwarfs the possible impact of sea level rise.

A rise in sea level also would increase the salinity of rivers and estuaries. Since the last ice age, as sea level rose approximately one hundred meters (several hundred feet), such freshwater rivers as the Susquehanna have evolved into estuaries like the Chesapeake Bay. A decrease in the flow of a river or an increase in the volume of water allows salt to migrate upstream. An increase in sea level of only thirteen centimeters (five inches) could result in salt concentrations in the Delaware River migrating two to four kilometers (one to two miles) upstream (Hull and Tortoriello, 1979). A rise of one meter could cause salt concentrations to migrate over twenty kilometers, possibly enough to threaten part of Philadelphia's water supply during a drought. Because some rivers recharge aquifers, some aquifers might become salty as well.

The impacts of storm damage and salt intrusion may be exacerbated or mitigated by the impacts of an enhanced greenhouse effect on climate. For example, possible increases in hurricane frequencies would further increase storm damage, while reductions in the severities of northeasters could reduce it. Flohn (1981) has suggested that in the mid-latitudes less precipitation might result, which would amplify salinity increases. Because these possibilities are still very tentative, they are not included in the analyses presented here.

Impacts on Today's Decisions

The costs associated with the physical effects of sea level rise could be very high. Although the worst effects would not begin to be felt until

2025, low-lying areas and beach resorts could be seriously affected before then. Furthermore, a wide variety of decisions made in the next decade will significantly influence the extent of the damages from sea level rise in the next century.

A popular convention used in evaluating large-scale projects is to assume that the project has a lifetime of thirty years. For many purposes, this convention is useful. Unfortunately, it has also led some managers to view the future beyond thirty years hence as completely irrelevant, even for projects that last much longer. Although machinery may only last ten years, many factories last fifty to one hundred years. Although pavement may last only ten years, a road lasts and channels land development for centuries. The location and layout of most major cities in the eastern United States were determined by decisions made before 1800. Houses, bridges, port facilities, airports, utilities, cathedrals, and office buildings constructed in the next decade may be useful for the next century or longer.

Sea level rise could affect all of these projects. Buildings could be destroyed by erosion and storms. Federal government programs that aid victims of natural disasters could become much more costly. Roads could be destroyed, and the costs to localities of maintaining infrastructures could increase. Bridges, docks, and aids to navigation would have to be reconstructed. Communities with high water tables would have to redesign drainage systems, and basement flooding could become more severe. Salt intrusion could necessitate constructing expensive desalinization facilities or relocating water intakes. Existing riparian rights and pacts to distribute water between municipalities might become unfair.

Environmental Impacts of Sea Level Rise

Like the physical effects, the environmental impacts of sea level rise fall into the categories of shoreline retreat, salt intrusion, and increased flooding. Perhaps the most serious environmental consequence would be the inundation and erosion of thousands of square miles of marshes and other wetlands. Wetlands (areas that are flooded by tides at least once every 15 days) are critical to the reproductive cycles of many marine species. Because marsh vegetation can collect sediment and build upon itself, marshes can "grow" with small rises in sea level. But for faster rates of sea level rise, the vegetation will drown. Its resulting deterioration may significantly erode land previously held together only by the marsh vegetation. Relative sea level rise of one meter per century is eroding over one hundred square kilometers (about fifty square miles) per year of marshland in Louisiana.

Salt intrusion is a threat to marine animals as well as vegetation. Many species must swim into fresher water during reproduction. In response to sea level rise, fish might swim farther upstream, but water pollution could prevent such an adaptation from succeeding. Some species, on the other hand, require salty water, such as the oyster drill and other predators of oysters. Consequently, salinity increases have been cited for the long-term drop in oyster production in the Delaware Bay (U.S. Fish and Wildlife Service, 1979; Haskin and Tweed, 1976), as well as recent drops in the Chesapeake Bay. Salt intrusion could also be a serious problem for the Everglades.

Flooding could have a particularly important impact on environmental protection activities. As Chapter 9 indicates, regulations for hazardous waste sites promulgated under the Resource Conservation and Recovery Act currently impose special requirements for sites in 100-year flood zones. Another EPA program, Superfund, has responsibility for abandoned waste sites, some of which are in low-lying areas such as Louisiana and Florida that could be inundated.

There are over one thousand active hazardous waste facilities in the United States located in 100-year floodplains (Development Planning and Research Associates, 1982) and perhaps as many inactive sites. Sea level rise could increase the risk of flooding in these hazardous waste sites. For example, if a hazardous waste facility is subjected to overwash by strong waves or simply to flooding that weakens the facility's cap, the wastes can be spread to nearby areas, thus exposing the population to possibly contaminated surface water. Moreover, by intruding into clay soils (which are often used as liners for hazardous waste disposal) saltwater can increase leaching of wastes.

RESPONSES TO SEA LEVEL RISE

We briefly review here the numerous methods that are available to prevent, mitigate, and respond to erosion, flooding, and salt intrusion from sea level rise. Sorensen et al. provide more detail in Chapter 6.

Communities and individuals must decide whether to attempt to protect themselves from the consequences of sea level rise or adapt to them. Generally, prevention will be economically justifiable only at valuable locations, such as population centers, defense installations, historical sites, and areas of environmental importance such as habitats of endangered species. Other areas would have to adjust to the consequences.

Prevention of erosion requires keeping waves from attacking the shore. This is generally achieved by intercepting the waves offshore or by armoring the beach itself. Offshore breakwaters limit the size of incoming waves. Revetments armor the beach itself and can be useful for moderate size waves.

Several means of preventing inundation and storm surge also serve to limit erosion. Seawalls, levees, and bulkheads are vertical wall structures made of materials of various strengths, depending on the size of the waves. New Orleans and other low-lying communities are protected by levees, while Galveston is protected by a seawall. With a rising sea, however, these structures may require protection themselves. Shoreline retreat in Galveston, for example, threatens the seawall's foundations. Accurate forecasts of future sea level rise could enable engineers to determine the heights and best design of these structures so that their initial construction is appropriate and cost-effective for their entire lifetime.

Because beaches and waves are important to resort communities, structures are not always an acceptable response to erosion, inundation, and storm surge. A popular but expensive option is artificial beach nourishment, that is, pumping sand from offshore or dredging a nearby channel. Because it would increase the amount of sand required, a rise in sea level could significantly increase the cost of such activities. In Chapter 8, however, Titus argues that the substantial real estate values would justify beach nourishment in many resorts, provided the sand was available.

Restoring other mechanisms of natural systems can also protect against erosion and storm surge. For example, dunes can provide a reservoir of sand to slow erosion and act as a levee against storm surges. Also, some marshes are supplied with sufficient sediment during floods to keep up with sea level rise. Where dunes have been destroyed or rivers leveed to prevent flooding, restoring these natural mechanisms may be cost effective.

Adjustment to the physical consequences of sea level rise may sometimes be more appropriate than prevention. In anticipation of erosion, some communities may prohibit construction in the most hazardous areas, and abandonment may even be necessary. In Chapter 8, Titus suggests that in the aftermath of a devastating storm, low-density communities might require development to retreat landward by fifty meters (one to two hundred feet). Such a policy could prevent subsequent losses to erosion and storms and help preserve a recreational beach. In the case of barrier islands, he recommends that communities consider imitating natural overwash processes by pumping sand to the bay side to preserve total acreage. Marsh systems could be maintained by identifying and reserving higher ground for migration, and later, by planting marsh vegetation.

Communities could adapt to increased storm damage by using measures already required in many hazard-prone areas. Houses can be elevated on pilings, waterproofed, and designed so that the first floor is a carport or utility area. Orienting structural walls parallel to the prevailing wave direction can protect them from destruction by storm waves. Commercial buildings can be designed so that the most valuable equipment is above future flood levels.

Adaptation to erosion and storm damage requires more advanced warning of sea level rise than building protective structures. For example, once completed, a building frequently is used for fifty to one hundred years; a highway influences development even longer. In contrast, protective structures and beach restoration can be accomplished in only a few years.

As with erosion, inundation, and flooding, individuals may either prevent or adapt to salt intrusion. In rivers, salt migrates upstream from both sea level rise and droughts. Therefore, preventive methods that currently focus on droughts could be extended to incorporate sea level rise. The Delaware River Basin Commission has responded to salt intrusion by constructing reservoirs that release water during droughts, maintaining a minimum flow. Areas that rely on rivers for drinking-water supplies can also maintain the flow by restricting consumption during droughts. Smaller communities can respond by moving intakes upstream or shifting to alternative supplies.

Most marine species can respond to salt intrusion by migrating upstream. Although sessile species such as oysters cannot move upstream fast enough to respond to salinity increases from droughts, the gradual rise in sea level would probably be slow enough to accommodate a migration. Because water pollution from urban areas upstream might make such a migration impractical, additional water pollution controls might be necessary.

The most frequent response to salt intrusion into a coastal aquifer is to seek alternative water supplies, such as wells farther inland. However, valuable aquifers, such as the Potomac-Raritan-Magothy aquifer system in southern New Jersey, might warrant engineering solutions. Freshwater can be injected near the salty body of water that is recharging the aquifer, forming an injection barrier that reverses the flow of water back into the saltwater body. Extraction of the intruding salt water, physical barriers, and increasing the amount of freshwater available to recharge the aquifer are other options. However, all of these options are expensive and have had only limited application.

The increased risk of flooding hazardous waste sites could be addressed by strengthening existing programs, particularly as they apply to closed and abandoned sites. As Chapter 9 discusses, EPA regulations already require operating waste sites in 100-year floodplains to ensure that wastes do not escape and contaminate surrounding areas during floods. With a rise in sea level, the 100-year flood boundary would shift inland, and these regulations would require more sites to undertake flood mitigation measures as the risks increased. However, existing regulations provide no similar protection against contamination from closed or abandoned sites. Regulations governing the closure of waste sites in the future could be modified to ensure that the sites are secure in the event of sea level rise.

METHODS USED IN THE CASE STUDIES

The case studies were innovative approaches to problems that previously had not been explored. Because in many instances there was little or no research on which to build, we adopted the case study approach so that our efforts would produce methods as well as results.

Each analysis required inputs from the previous analysis. Using the projections of sea level rise, Kana et al. and Leatherman projected the shoreline retreat and storm surge that would result if no additional protective measures were implemented. Using this information, Gibbs projected the economic impacts of sea level rise, both for the cases where sea level rise is and is not anticipated. In both cases, he used the analysis of Sorensen et al. to develop possible structural responses to sea level rise. The difference between these impacts (i.e., with and without anticipation of sea level rise) provides a measure of the value of policies that anticipate sea level rise. Titus used Gibbs's estimates of economic impacts to explore homeowners' decisions of whether to rebuild oceanfront houses destroyed by a major storm.

Choices of Study Areas

Several factors were considered in choosing our case study sites. We wanted to represent different coasts and different tidal and erosional patterns. We wanted commercial and industrial development patterns to vary. The costs of obtaining data covering both natural phenomena (such as the National Weather Service's Storm Surge Model) and socioeconomic variables had to be within the study's budget. The availability of expert coastal scientists with extensive experience in that particular study area was also considered important.

In response to these considerations, Galveston and Charleston were chosen as study areas. Galveston's history of subsidence and the availability of maps dating back to 1850 provided a record of the impact of relative sea level rise on the area. The two areas have different tidal patterns, Galveston Bay being *microtidal* (tide ranges average less than two feet) and Charleston being *mesotidal* (tide ranges average five to six feet).

Charleston and Galveston also exhibit different industrial development, resources, amenities, and protective approaches to storm threats. The most highly developed part of the Galveston study area is directly exposed to the ocean, with extensive protective structures throughout the area. Charleston lies behind a string of barrier islands and has few coastal works other than a seawall guarding the tip of the peninsula. Charleston's extensive historic district poses special economic and environmental challenges, while Texas City, in the Galveston study area, boasts one of the country's largest petrochemical and refinery complexes. Growth within the Galveston area is limited by land subsidence and groundwater shortages, while parts of Charleston will experience rapid growth over the next two decades. The National Weather Service's new Storm Surge Model was available for the Galveston area. Finally, Leatherman (for Galveston) and Kana et al. (for Charleston) had extensive experience in coastal research and mapping of their respective areas.

Projecting Shoreline Retreat

Sea level rise causes shorelines to retreat both because land lying below future sea level will be permanently inundated and because erosion of nearby land will increase. The particular method appropriate for estimating shoreline retreat at given points on the coastline depends upon topography, beach composition, wave climate, sediment supply, and available historical data.

A theoretical model for estimating the impact of sea level rise on shoreline retreat is provided by the Bruun Rule (see Figure 1-3). This rule assumes that after a rise in sea level, the beach profile that existed prior to the rise will be restored through wave action eroding away the upper part of the beach and redepositing the material on the bottom. Essentially, the Bruun Rule says that shoreline retreat should be predicted using the average slope of the entire beach system from the dune crest to an area several thousand feet out to sea, rather than the slope of the portion of the beach immediately above sea level.

Despite its importance as a conceptual tool, the Bruun Rule is insufficient to predict shoreline retreat. If a certain percentage of sediment is likely to be carried away, the method must be adjusted by using an estimate of the percentage of material lost. Furthermore, estimating the offshore limit of the beach system can be difficult and involves an element of judgment. Finally, the Bruun Rule only estimates shoreline changes caused by sea level rise, while our analysis requires estimates of all factors influencing shoreline location. Therefore, even where the Bruun Rule can estimate shoreline retreat due to sea level rise, it may be necessary to rely on other methods to account for shoreline changes caused by other factors.

In the Galveston case study, Leatherman used an empirical model for determining shoreline retreat and concluded that the model was consistent with results obtained from applying the Bruun Rule. Using maps dating back to 1850, he determined that the local relative sea level rise (global sea level rise plus subsidence, which has been of major importance in the Galveston area) of forty centimeters (sixteen inches) was the only major cause of the shoreline retreat that had been observed. Because the area has a constant slope, he assumed that another forty centimeter rise would result in the same amount of shoreline retreat as had been observed in the past. Therefore, Leatherman predicted shoreline retreat by an empirical formula that says, essentially, that each centimeter of future sea level rise will result in a shoreline retreat equal to one-fortieth the retreat that has occurred since 1850.

Determining the impact of sea level rise on the Charleston area presented a more difficult problem. First, the record for shoreline change was available only for the last 40 years, during which relative sea level rose only ten centimeters (four inches). Equally important, because the three rivers that converge to form Charleston Harbor deposit significant amounts of sediment and much of the shore had actually advanced, historical sea level rise has been only one of many factors responsible for historical shoreline change. Furthermore, because Charleston Harbor is narrower than Galveston Bay, the waves are smaller. Therefore, wave-induced erosion (predicted by the Bruun Rule) would not be as significant. On the other hand, a rise in sea level would slow river currents and alter the amount of resulting shoreline change. Finally, while most of the case study area was in the harbor, the area also included a barrier island (Sullivans Island).

Kana et al. generated a baseline shoreline by extrapolating past trends, making allowances for the probable rediversion of the Santee River, which would reduce sediment supply. Their projections of shoreline change due to sea level rise within the harbor assumed that all shoreline changes would be due to inundation (i.e., no erosion would result from sea level rise within the harbor). For Sullivans Island, Kana et al. used the Bruun Rule to predict erosion due to sea level rise until the island reached a critical width. At that point, they assumed that the island would migrate landward at a rate of six meters per year, on the basis of experience with other barrier islands in "overwash mode" in the region. All existing development on the island would be destroyed as the island migrated by approximately its own width.

For protected shorelines, both case studies assumed that the protective structures would halt all erosion up to the point where they were overtopped. Leatherman, however, points out that earthen levees in Texas City would erode before being overtopped unless they were reinforced.

Storm Surges

Storm surge refers to the superelevation of water associated with hurricanes and northeasters. Predictions of storm surge elevations are generally based on historical records of previous storms. The Galveston Bay area was one of the first areas modeled by the National Hurricane Center using the SLOSH (Sea, Lake, and Overland Surges for Hurricanes) model, which Leatherman used to estimate existing storm surge frequency. SLOSH simulates wind speeds and storm surges based on the probabilities of various combinations of tides, meteorological conditions, topography above and below the water, and existing coastal structures. The model estimates the frequency of flooding and maximum surge. Because this model was not yet available for Charleston, a previous analysis based on the National Oceanic and Atmospheric Administration's SPLASH (Special Program to List Amplitudes of Surges from Hurricanes) model was used to predict storm surge frequencies and magnitudes given existing sea level.

Both case studies estimated the new storm surge levels for areas already in flood zones by adding the amount of sea level rise to the amount of flooding predicted by SPLASH and SLOSH under current conditions. Both assumed that the floodplain boundaries would move inland to the point where the resulting increase in elevation of the boundaries was equal to the rise in sea level. (This assumption of no attenuation of the flood surge would not be appropriate for very flat areas, such as Florida and the Mississippi Delta.)

For protected areas, Kana et al. and Leatherman assumed that there would be no flooding unless surges were great enough to overtop seawalls. Although sea level rise would subject some barriers to greater stresses than they were designed to withstand, Kana et al. and Leatherman assumed they would remain intact. Both assumed that once a barrier was overtopped, the water level on the protected side would rise to the level to which it would have risen without the seawall. Although a barrier should provide some protection, Leatherman believes this assumption to be reasonable for Texas City because of the city's small size. In the case of Galveston, flooding would occur from the bay side before the seawall was overtopped.

Salt Intrusion

The two case studies considered only the salt intrusion into aquifers, not surface waters. The "Ghyben-Herzberg relation" was used to estimate the present location of the freshwater/saltwater boundary. This principle states that the depth of the saltwater/freshwater interface is forty times the elevation of the water table above mean sea level. This is conservative in that the boundary has undoubtedly moved landward due to over-pumping. Sea level rise was assumed to shift the water table and freshwater/saltwater boundary upward by the amount of sea level rise and landward in accordance with shoreline retreat.

Admittedly, more sophisticated models might have been used. However, we did not believe that salt intrusion into aquifers warranted additional investigation because the salt intrusion from overpumping in the Charleston area dwarfed all impacts from sea level rise, and Galveston-Harris County prohibits additional pumping of groundwater because of historical problems with subsidence.

Economic Impacts

Gibbs's economic analysis in Chapter 7 proceeds in two steps. First, he defines and measures the economic value of the land affected by shoreline movement and storm surge. The economic impact estimates are measured in terms of the real resource costs to society caused by sea level rise. Then, he analyzes the value of anticipating sea level rise. Gibbs's analysis does not, however, consider the impacts of salt intrusion or the impacts (positive or negative) on parties outside the study area.

Real economic losses fall into three categories: (1) the direct losses of economic services from land and capital caused by shoreline retreat and storm damage; (2) the cost of protection, mitigation, and response measures taken to reduce these losses; and (3) lost opportunities due to sea level rise. Gibbs encompassed these three consequences with a single measure called "net economic services." This measure is the value of all services produced minus the costs of producing them (costs include expenditures for new investment, maintenance, and protection and mitigation actions). Because structures remaining at the end of the time period analyzed will continue to produce economic services, their value must also be considered.

To compute net economic services, Gibbs simulated investment, the damages from storms and erosion, and prevention, mitigation, and response measures for each decade. Because human behavior is difficult to predict, Gibbs examined the sensitivity of the economic analysis to various parameters and assumptions. For example, he varied the behavioral assumptions that determine development patterns and the choice of protective actions, with different responses to the same experience and information being tested.

Expectations will play a key role in determining future damage. By considering behavioral responses, the analysis explicitly accounted for the effects of expectations of sea level rise on future decisions. For example, if no one anticipates sea level rise, certain areas could be developed, only later to face the threats of shoreline retreat and storm damage. In this instance, the costs of storm damage (and possibly of protective measures needed later) would increase. If sea level rise were anticipated, however, such areas might be developed differently or not at all, reducing adverse impacts. Three types of community response action are used in the analysis: stop or reduce the rate of shoreline movement through the use of revetments, levees, or other means; eliminate the threat of storm surge (up to a given elevation) through the use of seawalls and levees; and reduce or prohibit investment in given areas by promulgating land use regulations.

In computing net economic services, the lost opportunities associated with less development and the cost of building protective structures were subtracted.

The Value of Anticipating Sea Level Rise

The value of anticipating sea level rise and its consequence depends on how much people change their behavior to avoid the resulting economic losses. Gibbs estimated this value using a variety of assumptions about how people would change their behavior. Changes in both private investment and community planning were considered. Private investment was simulated to be reduced in areas of increasing hazard due to sea level rise. Changes considered in community actions include forgoing development of areas that would be lost because of sea level rise and taking more timely and effective protective measures.

Gibbs did not examine impacts on environmental amenities. Later analyses will need to consider these impacts. For example, sea level rise could destroy the marsh habitat of an endangered species, and advanced planning could save the marsh. The value of saving the marsh would include such disparate consequences as savings to the fishing industry and preventing a species from becoming extinct.

CASE STUDY RESULTS

This section summarizes the impacts of sea level rise on the Charleston and Galveston areas. The physical impacts of sea level rise are summarized in terms of the area of land lost and changes in the areas subject to flooding. (Chapters 4 and 5 present detailed maps showing these effects.) Because salt intrusion into groundwater from sea level rise was not projected to be significant, this impact is discussed only briefly.

Finally, we summarize Gibbs's estimates of the economic impacts of sea level rise and the extent to which these impacts could be reduced by policies that anticipate this rise.

Before discussing the results of the case studies, we strongly emphasize that these results should be viewed with extreme caution, particularly the projections for specific neighborhoods. The case study Chapters 4, 5, 7, and 8 are initial applications of methods developed for this book. Although the methods use realistic assumptions and rely as much as possible on empirical evidence, none of the authors can be certain that all major factors were adequately considered. Therefore, the results should be viewed as approximations to illuminate our understanding of sea level rise, not as precise forecasts of the fates of particular city blocks. This caution applies most of all to the Chapter 8 analysis of Sullivans Island. We also remind the reader that we expect the actual rise in sea level to be between the low and medium scenarios. Although we investigated the high scenario, we believe it to be very unlikely.

Charleston Study Area

Description. The Charleston study area consists of the land around Charleston Harbor, which is formed by the confluence of the Cooper, Ashley, and Wando Rivers (see Figure 1-4). The study area includes all of Charleston and parts of North Charleston, Mount Pleasant, Sullivans Island, and James Island. The shores of the harbor are dominated by fringing salt marshes and tidal creeks. Lower Charleston peninsula, in the center of the study area, has a maximum elevation of only six meters (eighteen feet) and includes several low-lying areas that have been reclaimed from the harbor. North Charleston, on the upper part of the peninsula, has elevations up to ten meters (thirty feet). West Ashley, to the west of the peninsula, is a relatively flat area fronted by extensive marshes along the shores of the harbor and the tidal creeks, with elevations of three meters or less. Mount Pleasant, while also flat, is generally higher, with elevations between three and ten meters. Sullivans Island, in the northeast portion of the study area, is a narrow barrier island whose average elevation is less than three meters above sea level.

Because of the harbor's funnel shape, tides range up to two meters (about six feet). Although the Charleston area does not have a history of extensive hurricane damage, the tides and the extensive network of tidal creeks expose parts of the peninsula, West Ashley, and Sullivans Island to periodic flooding. The Cooper River has recently been responsible for a large amount of sedimentation, which has led to the accretion of shorelines in the marshy areas within the harbor.

The only major protective structure in Charleston is the Battery, a six foot seawall located at the tip of Charleston Peninsula. Even today, a 100-year storm would overtop the Battery.

The Charleston study area had a population of approximately 120,000 in 1980. Charleston Peninsula, with over 70,000 people, is the economic and population center for the study area. The southern end of the peninsula has a densely populated historic district and other residential, commercial, and port areas; the central peninsula consists of industrial parks and marshland; and the upper peninsula (North Charleston) has a combination of residential areas and heavy industry, including a very large naval reservation.

Because most of the peninsula is already highly developed, the potential for the Charleston area to grow is limited. Undeveloped land is scarce, and although some growth may take place in the northern portion or elsewhere through shifts to more high-density land uses, the long-term growth rate for population and employment has been estimated at 0.8 percent per year for the next fifty years.

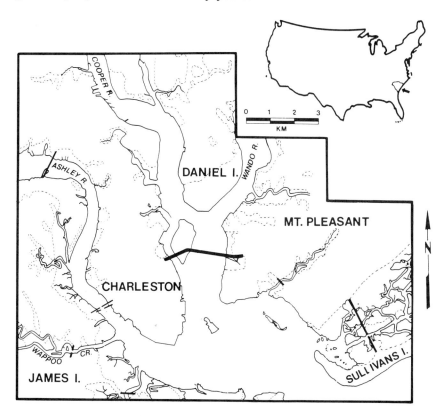

Figure 1-4. Charleston study area.

West Ashley and James Island (population 8,500 within the study area), on the mainland to the west of the peninsula, consist mostly of low-density single-family housing. Portions of these communities lie within the city limits of Charleston. Future development will be mostly single-family housing on currently vacant land.

Mount Pleasant (population 21,800) is a residential commuter town with attendant commercial development. It has the greatest potential for growth of any community in the study area. Most of this growth will be in the form of additional housing, but extensive industrial development will probably take place to the north, near the new South Carolina Port Authority terminal and the planned expressway.

Sullivans Island (population 10,000) is a residential and resort community located to the east of Mount Pleasant along the coast. The island itself has been extensively developed with single-family homes, with high-rise and condominium construction currently prohibited by zoning regulations. Changes in those regulations would be a prerequisite to any substantial growth on Sullivans Island. Because of sediment supply and a jetty that protects Charleston Harbor, much of the island is currently accreting.

The Charleston study area has five hazardous waste facilities in the current 100-year floodplain, and six outside the floodplain. Of the hazardous wastes stored, treated, or disposed of at these sites, carcinogens and ecotoxins probably present the greatest risks to human health and the environment in the event of a release. The types of hazardous wastes at these facilities include cadmium, arsenic, benzene, beryllium, chromium VI, nickel, and vinyl chloride.

Impacts of Sea Level Rise. The analysis by Kana et al. concludes that up to one-half of the Charleston area could be permanently flooded if no response actions were taken. Gibbs concluded that taking anticipatory actions could save the area as much as $1.5 billion.

Table 1-4 summarizes the impacts of sea level rise on shoreline retreat and on the 10- and 100-year floodplains for the years 2025 and 2075 under the low, medium, and high scenarios, as well as an extrapolation of current trends. In general, the impacts of the high scenario in 2025 are slightly less than the impacts of the low scenario in 2075. Even if only current trends continued, the study area would lose 4 percent of its land by 2075, mostly on Sullivans Island and in the marshes along Charleston Harbor. Under the low scenario, 5.2 percent of the land would be lost by 2025 and 15 percent by 2075. Under the high scenario, 14 percent of the area would be lost by 2025 and 45 percent by 2075.

Tables 1-A through 1-C in the appendix provide similar estimates of the impacts on specific communities. In the medium scenario, Sullivans Island could lose the first one or two rows of houses along the ocean by

Table 1-4. Charleston Study Area: Summary of Direct Physical Impacts by Scenario (in km², percent of total area given in parentheses)

Scenario	Year	Area Lost because of Shoreline Movement	Area in 10-Year Floodplain[a]	Area in 100-Year Floodplain[a]
No Sea				
Level Rise	1980	_[b]	30.8 (32.9)	59.2 (63.2)
Trend	2025	1.8 (1.9)	32.9 (35.1)	61.1 (65.2)
Scenario	2075	3.9 (4.2)	34.9 (37.2)	62.9 (67.1)
Low	2025	4.9 (5.2)	35.7 (38.1)	63.7 (68.0)
Scenario	2075	14.2 (15.1)	45.0 (48.0)	71.2 (76.0)
Medium	2025	7.8 (8.3)	38.6 (41.2)	66.0 (70.4)
Scenario	2075	28.7 (30.6)	58.5 (62.4)	78.7 (84.0)
High	2025	13.0 (13.9)	41.4 (44.2)	68.4 (73.0)
Scenario	2075	43.0 (45.9)	69.4 (74.1)	83.9 (89.5)

Note: One square kilometer equals 0.38 square miles.
[a] Includes area lost because of shoreline movement.
[b] Total area in 1980 is 275 sq km.

2025 and would migrate landward by its own width by 2075, destroying virtually all existing development. Developed portions of Charleston Peninsula would not be threatened under the low scenario, partly because of existing seawalls and bulkheads. However, these protective structures would not prevent inundation of one-quarter and one-half of the peninsula by 2075 for the medium and high scenarios, respectively. The West Ashley/James Island area would be even more vulnerable, and Mount Pleasant would be the least affected.

Table 1-4 also shows that by 2075 in the medium scenario, a 10-year storm would cause as much flooding as a 100-year storm would inflict today. About one-third of the study area is currently in the 10-year floodplain, and about two-thirds is within the 100-year floodplain. By 2075 under the low scenario, almost one-half the study area would be in the 10-year floodplain, and three-quarters within the 100-year floodplain. Under the high scenario, 75 percent of the study area would be within the 10-year floodplain by 2075, and almost 90 percent of the study area, including five additional hazardous waste sites, would be in the 100-year floodplain.

Kana et al. found that the freshwater/saltwater interface could shift up to sixty meters (two hundred feet). They concluded that this impact would be negligible, compared with the impact of overpumping on salt intrusion.

Shoreline retreat and additional storm flooding could inflict heavy economic losses. Gibbs estimated the economic impacts of sea level rise for two assumptions about how individuals and communities would address sea level rise: (1) people would have no foresight and would

Table 1-5. Charleston Study Area: Summary of Economic Impacts by Scenario

Scenario	Years	Economic Impact If Sea Level Rise Is Not Anticipated (% of Total Economic Activity)	Economic Impact If Sea Level Rise Is Anticipated	Value of Anticipating Sea Level Rise (% of Economic Impact)
Low	1980–2025	280 (4.9)	160	120 (43)
	1980–2075	1,250 (17.3)	440	810 (65)
Medium	1980–2025	685 (12.0)	345	340 (50)
	1980–2075	1,910 (26.5)	730	1,180 (62)
High	1980–2025	1,065 (18.7)	420	645 (60)
	1980–2075	2,510 (34.8)	1,110	1,400 (56)

Note: All values are in millions of 1980 dollars valuated at a real discount rate of 3 percent per year.

adapt only in response to the observed effects of sea level rise; and (2) people would use foresight to adapt in anticipation of these impacts. The value of anticipating sea level rise would be the difference between these two impacts. Table 1-5 displays Gibbs's estimates of economic impact for each of the sea level rise scenarios. This table combines results shown for storm surge and erosion damages in Table 1-C in the appendix. The assumptions used to calculate these results are presented in detail in Chapter 7.

The cumulative economic impact in the Charleston study area would range from $280 million in the low scenario through 2025 to $2.5 billion in the high scenario through 2075, if sea level rise were not anticipated.[12] These impacts range from 5 percent to 35 percent of the total economic activity that would take place in the study area in the absence of sea level rise.

Gibbs concluded that the impacts could be reduced by 43-65 percent by anticipating sea level rise. Anticipation of sea level rise was represented by reducing private investment in areas of increasing hazard and more timely implementation of community responses, such as the construction of seawalls and levees.

Although Gibbs assumes that most options would be implemented after 2000, Titus concludes in Chapter 8 that by 1990 sea level rise may be a critical issue to Sullivans Island, a barrier resort. His conclusion was based on the data underlying Kana's projection that the first or second row of houses could be eroded by 2025 under the high scenario and on Gibbs's unreported result that a 100-year storm would devastate much of the island. Titus concludes that unless the community plans to pump increasing amounts of sand onto the beach for the foreseeable future, perhaps 20 percent of the houses should not be rebuilt in their original locations after a storm in 1990 if the high scenario is expected.

Galveston Study Area

Description. The Galveston study area includes the northern third of Galveston Island, the top of Bolivar Peninsula, and the nearby mainland areas of Texas City, La Marque, and San Leon (see Figure 1-5). The land throughout the study area is primarily a gently sloping coastal plain broken by estuaries and lagoons along the shores of Galveston Bay. Most of the study area is less than five meters above sea level. Tides

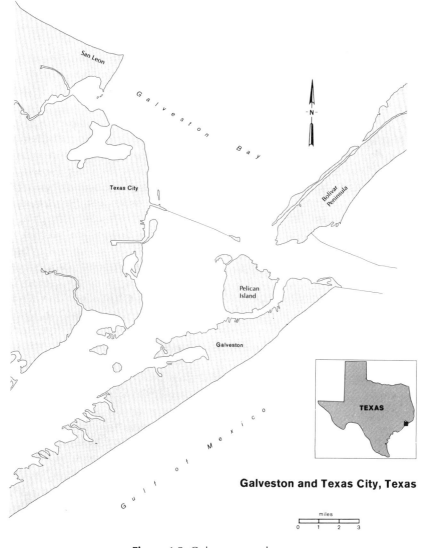

Galveston and Texas City, Texas

Figure 1-5. Galveston study area.

in the area range from fifteen centimeters in Galveston Bay to sixty centimeters in the Bolivar Road inlet.

The Galveston study area has a history of shoreline retreat and storm damage. Six thousand people died in the 1900 hurricane, the worst natural disaster in U.S. history. There has been a considerable amount of land subsidence within the study area over the past century, causing a relative sea level rise of more than thirty centimeters (one foot) along the coast. Historical records show that this sea level rise has been accompanied by shoreline retreat throughout the study area. Consequently, communities in the area have built a variety of structures to reduce erosion and flooding of developed areas, including seawalls, levees, and pumping facilities.

The Galveston study area had a population of approximately 122,000 in 1980. The area is expected to grow moderately in the future with La Marque, Texas City, and San Leon growing much faster than Galveston. Major commercial development should occur in Texas City, just to the west of the study area. A lack of water supplies may, however, impede both industrial and residential development in parts of the mainland.

The city of Galveston (population 65,000) is located on the northern third of Galveston Island. This portion of the study area is mostly developed and includes residential housing, commerical districts, light industrial and port facilities, and the University of Texas medical center. A five meter seawall runs along the south side of Galveston, protecting it from storm waves and gulfside flooding. The downtown section of Galveston has sufficient elevation to avoid flooding from the bayside. However, other developed parts of the city experience flooding even during a 15-year storm.

There is little room for further development on that part of the island within the study area. Galveston's economy, based on shipping, transportation, medicine, tourism, and recreation, has limited long-term development potential, and the population of this part of the study area is likely to remain stable or increase slightly over the next fifty years. Pelican Island consists of marsh and dredge spoils, with some university and shipping facilities. The part of Bolivar Peninsula in the study area includes only a few hundred houses.

Texas City and La Marque constitute the geographic and economic center of the study area, with a population of 57,000 and taxable property valued at over two billion dollars. Three-quarters of Texas City consists of undeveloped, low-lying floodplain, some of which has been further affected by land subsidence. That portion of La Marque within the study area has undeveloped marshes to the south and southeast, and low-density residential areas and attendant commercial development in the center of the city. The densely developed portions of Texas City and La Marque are protected from storm surge by an extensive network of structures, including the Texas City Levee System. However, a 100-year storm would currently cause over $130 million in damage to the unprotected, moderately developed areas.

Texas City and La Marque's economy is based on petrochemicals, petroleum refining, shipping, and land transport. One-half of the developed land is occupied by a one billion dollar petrochemical complex, which provides the major employment and tax base for the region. The continued strength of the energy and energy-related sectors should cause these communities to grow more rapidly than Galveston and increase their population by one-third by the end of this century.

San Leon is a residential area for commuters who work in Houston, Galveston, and Texas City. There is little commercial development in this area. Its population of 2,000 is expected to double over the next twenty years. In spite of a history of shoreline retreat, San Leon has little protection against erosion or storm surge.

In the Galveston study area, the 100-year floodplain contains ten hazardous waste facilities. The carcinogens identified to be located at these sites include benzene, carbon tetrachloride, chromium VI, polynuclear aromatic hydrocarbons, beryllium, nickel, cadmium, and arsenic. The ecotoxins identified include the pesticides methyl parathion and lindane.

Impacts of Sea Level Rise. Because Galveston and Texas City are largely protected by seawalls and levees, the impact of sea level rise would not be as great for this study area as for Charleston. However, by 2075, a 100-year storm would overtop the Galveston seawall in the medium and high scenarios. Damage from such a storm would be approximately two billion dollars—four times greater than if sea level did not rise.

Table 1-6 shows the area lost to sea level rise for the four scenarios. Current sea level trends would erode about 2.5 percent of the study area by 2075. The low scenario would result in a loss of 1.5 and 6 percent of the study area by 2025 and 2075, respectively. Under the high scenario, 3 and 12 percent would be lost by 2025 and 2075, respectively. With the exception of Sea Leon, whose entire peninsula would erode under the high scenario, the erosion would take place in undeveloped areas, the only places not protected by seawalls.

The impact of sea level rise on floodplains would be more significant. As Table 1-6 shows, one-quarter of the study area is now in the 15-year floodplain. By 2075, this proportion would increase to one-third under the low scenario and over one-half in the high scenario.

In particular, up to 80 percent of Galveston would be vulnerable to a

Table 1-6. Galveston Study Area: Summary of Direct Physical Impacts by Scenario (in km², percent of total area given in parentheses)

Scenario	Year	Area Lost because of Shoreline Movement	Area in 15-Year Floodplain[a]	Area in 100-Year Floodplain[a]
No Sea				
Level Rise	1980	_[b]	65.3 (23.7)	160.6 (58.4)
Trend	2025	2.6 (0.9)	71.7 (26.1)	163.2 (59.3)
Scenario	2075	6.2 (2.2)	77.4 (28.1)	165.2 (60.1)
Low	2025	4.1 (1.5)	78.2 (28.4)	165.5 (60.2)
Scenario	2075	15.6 (5.7)	91.9 (33.4)	258.7 (94.1)
Medium	2025	6.5 (2.4)	82.9 (30.1)	167.0 (60.7)
Scenario	2075	24.4 (8.9)	119.7 (43.5)	267.3 (97.2)
High	2025	8.3 (3.0)	86.5 (31.4)	206.9 (75.2)
Scenario	2075	32.4 (11.8)	142.7 (51.9)	269.1 (97.8)

Note: One square kilometer equals 0.38 square miles.
[a] Includes area lost by shoreline movement.
[b] Total area in 1980 is 275 sq km.

15-year storm. By 2075 under the high scenario, a 15-year storm would inflict the amount of flooding that took place during hurricane Alicia in August 1983. The 10-year floodplain would increase from 60 percent of the study area currently to 95 percent by 2075, even in the low scenario. Under the medium and high scenarios, almost all the study area would be vulnerable, and storm waves would overtop all existing protective structures.

Twenty-two additional hazardous waste sites would be within the 100-year floodplain, for a total of thirty two under the high scenario for 2075. However, if the existing levees and seawalls were raised, these sites might not have to undertake any additional flood mitigation measures.

In their examination of the potential effects of sea level rise upon rates of salt intrusion into groundwater in the Galveston area, Leatherman et al. (1983) concluded that unconfined groundwater aquifers in the Galveston Bay area are generally polluted or salt-contaminated and that any incremental rise in sea level probably will have little effect on the two principal confined aquifers in the region.

Table 1-7 displays Gibbs's estimates of the economic impacts of sea level rise for two cases of adaptive behavior for Galveston. The actions he assumed would be taken in response to sea level rise are presented in detail in Chapter 7.

The cumulative economic impact in the Galveston study area ranges from $115 million in the low scenario through 2025 to $1.8 billion in the high scenario through 2075. These impacts range from 1.1 percent to 16 percent of the estimated total value of the economic activity that would

Table 1-7. Galveston Study Area: Summary of Economic Impacts by Scenario

Scenario	Years	Economic Impact If Sea Level Rise Is Not Anticipated (% of Total Economic Activity)	Economic Impact If Sea Level Rise Is Anticipated	Value of Anticipating Sea Level Rise (% of Economic Impact)
Low	1980–2025	115 (1.1)	80	25 (22)
	1980–2075	555 (4.9)	310	245 (44)
Medium	1980–2025	260 (2.6)	90	150 (58)
	1980–2075	965 (8.4)	415	550 (57)
High	1980–2025	360 (3.6)	140	220 (61)
	1980–2075	1,840 (16.0)	730	1,110 (60)

Note: All values are in millions of 1980 dollars valuated at a real discount rate of 3 percent per year.

take place in the study area over the same time periods in the absence of sea level rise. The economic value of damages would be less significant than in the Charleston area, given the smaller amounts of shoreline retreat and changes in floodplains.

The third column of Table 1-7 presents the savings from policies that anticipate sea level rise. Even in the low scenario, economic impacts can be reduced by over $245 million through 2075. Under the high scenario, impacts could be reduced by $220 million through 2025 and $1.1 billion through 2075.

Gibbs emphasizes that his methods are conservative and that the potential savings could be even greater. Chapter 7 presents estimates for both case study sites using alternate discount rates and discusses the sensitivity of the results to various assumptions about investment behavior and community responses to sea level rise.

REACTIONS AND RECOMMENDATIONS

The impacts of sea level rise on the Galveston and Charleston areas suggest that in the coming decades, sea level rise may become one of the most important issues facing coastal communities. Even today, erosion attributable to current trends is a major concern to Louisiana and many resorts. As Chapter 7 shows, many of the adverse consequences could be avoided if timely actions are taken in anticipation of sea level rise. Although some of these actions may not be necessary until 2000 and thereafter, others may only be timely if the planning process starts soon.

In March 1983, many of this book's findings were presented to a conference of over 150 scientists, engineers, and federal, state, and local policy makers. Although those attending agreed that sea level rise, if substantiated, would justify the attention of policy makers at all levels, some doubted whether anything less than a catastrophe could motivate people to undertake the necessary actions. Chapter 10 presents the reactions of six well-known representatives of the public and private sectors to our research and its implications.

Edward Schmeltz, an assistant vice president and department manager for coastal engineering at PRC-Harris, agrees with the conclusion of Chapter 6 that adequate technology is available to respond to sea level rise. He argues, however, that much greater confidence must be developed in the sea level rise projections before the engineering profession could convince clients to spend large sums of money to protect projects from sea level rise. Schmeltz argues that many people would view sea level rise projections as "hypothesis and conjecture." He further points out that many existing projects could withstand a one-half meter (two foot) rise in sea level but not a rise of three meters (ten feet).

Jeffrey Benoit, coastal geologist for the State of Massachusetts Coastal Zone Management Program, states that planning for sea level rise should start immediately if the projection of a four meter rise is correct. Like Schmeltz, however, he emphasizes that state agencies need a narrower range of uncertainty to address the rise properly. He also recommends that more attention be given to altered development patterns and regulation, in contrast to the "hard" coastal engineering responses described by Sorensen et al.

Sherwood Gagliano, who first popularized the relationship between relative sea level rise (subsidence) and coastal erosion in Louisiana, provides extensive comments on Chapters 4, 5, and 6. He concludes that the methods employed were very satisfactory for the sandy beaches of Galveston and Charleston but that future research should also consider the impacts on muddy beaches and changing tidal regimes.

Charles Fraser, chairman emeritus of Sea Pines Coropration (which developed Hilton Head Island, South Carolina) notes that institutions do not always respond to scientifically documented problems, even when the experts agree on the proper response. Furthermore, he questions whether coastal governments and property owners would be willing to consider the problems of the next century. He argues, however, that it could take several decades to develop societal responses and therefore that planning for sea level rise should continue.

Colonel Thomas Magness III (formerly assistant director for civil works, U.S. Army Corps of Engineers) notes that the Army Corps of Engineers has a planning horizon sufficiently long to prepare for sea level rise, and that the Corps is already starting to do so.

Lee Koppleman, executive director for the Long Island Regional Planning Board, indicates that on first reading, he thought that the prospect of sea level rise appeared to be sufficiently in the future that we might leave this issue for the next generation. He states, however, that as he thought about it, he decided that there is, in fact, a problem. Koppleman argues that planners can and will consider sea level rise if scientific research continues to be presented in a form they can understand. Gagliano also emphasizes the importance of presenting research in a useful form: "It was only after disclosure that a given coastal parish would last only 50 years before it eroded into the sea that the state legislature and the governor enacted a program for coastal erosion protection and shoreline restoration" (Chapter 10, page 300).

The reactions of the independent reviewers had two major messages in common: first, estimates of sea level rise must be improved; and second, even then, it will be difficult to induce an adequate response. The fact that many of the adverse consequences can be avoided does not guarantee that the necessary action will take place.

At this time, we can only speculate about the best way to overcome these difficulties. Because nothing will be done in the absence of information, increasing public awareness must be a top priority. Although this process will take time, researchers and professionals should not automatically assume that people will not plan for the future. At best, such an assumption ignores the substantial efforts that have been undertaken to respond to long-term problems such as population growth; at worst, the assumption could be self-fulfilling.

Nevertheless, it would be a mistake for research to focus only on the physical effects of sea level rise. We must also determine how to motivate society to act in a way that will lead it to be satisfied with the results of its actions, rather than regret its lack of foresight.

Based on the analysis presented in this book, the following recommendations are appropriate.

1. *Federal research on the physical, environmental, and economic impacts of sea level rise should be substantially expanded.* The pilot studies reported in this book provide rough estimates of some of the physical and economic impacts of sea level rise and of the value of preparing for these impacts. However, deciding which anticipatory measures should be implemented will require a better understanding of the impacts of sea level rise and possible responses.

The Army Corps of Engineers has already undertaken considerable research into the impacts of current sea level trends on beach erosion. That research should be expanded into a general model capable of predicting erosion from both storms and an accelerated rise in sea level.

However, the importance of such a model requires that experts in the private sector, academia, and other government agencies also participate in its development.

Government agencies charged with protecting the environment also must assess the vulnerability of their programs to sea level rise. For example, a one meter rise could devastate much of the existing wetlands in Louisiana and perhaps elsewhere. By undertaking the necessary research now, it may be possible to identify inexpensive ways to ensure that ecosystems and economic activities adapt to sea level rise without unnecessary conflicts.

2. *Federal support for scientific research on the rate of future global warming and sea level rise should be greatly expanded.* The benefits of this research would clearly justify the costs. Coastal communities could save billions of dollars by implementing timely actions in anticipation of sea level rise. But better forecasts of sea level rise will be necessary for these communities to take the right actions at the right time.

The highest priority should be research into the impact of a global warming on glaciers. Experts in glaciology could substantially improve upon the estimates of ice discharges used in this book, but have not been given the support necessary to adequately collect and analyze measurements and data produced by climate models. Other areas in need of research include the sources and sinks of the minor greenhouse gases, models of ocean currents, and the impact of a global warming on the frequencies, tracks, and severities of tropical storms and northeasters.

3. *State coastal programs should be strengthened.* Because of federal and state budget problems, many state coastal programs are being curtailed or eliminated. These programs are absolutely necessary to ensure that communities are provided with the required technical expertise and that adjacent jurisdictions adopt compatible response strategies.

4. *Federal, state, and local coastal programs should consider the impacts of accelerated sea level rise in their planning.* At the state and local levels, shore protection projects and post-disaster plans have a particular need to consider sea level rise. Communities should explicitly decide the amount of resources they are willing to invest to resist erosion. State and local governments that intend to maintain current shorelines should make the public aware of the ultimate cost of doing so.

Many federal agencies should also consider these impacts. The Federal Emergency Management Agency should consider the impact of sea level rise on its programs to prevent coastal flood disasters. The National Park Service and the Fish and Wildlife Service should consider whether their objective of maintaining marine ecosystems will require coastal uplands to be preserved so that future marshes can migrate landward. Finally, the

Corps of Engineers should consider the impact of sea level rise on its coastal engineering programs.

5. *Coastal engineers should revise standard engineering practices to consider accelerated sea level rise.* Coastal structures designed today will last well into the next century and perhaps longer, while soft engineering projects such as beach nourishment are very sensitive to even slight rises in sea level. Therefore, future sea level rise is likely to have an important impact on the outcome of coastal engineering decisions made today.

6. *Research into the most effective means of communicating risks and motivating effective responses should be undertaken.* Such efforts could draw on the Federal Emergency Management Agency's experience with individual and community responses to flood risks.

7. *A well-respected group of coastal engineers, planners, and other decision makers should conduct an independent review of the necessity of planning for sea level rise.* Practitioners cannot rely solely on the conclusions of researchers, whose incentives may differ from their own. Yet the individual engineer or planner will not have the time to review completely all of the evidence. A review panel could bridge the gap between researchers and practitioners.

This book discusses only the potential physical and economic impacts of sea level rise in the United States. However, the impacts could be much more serious in other parts of the world. In 1971, the storm surge from a tropical cyclone killed three hundred thousand people in Bangladesh. Countries in the Indian subcontinent, the eastern Mediterranean and other low-lying coastal areas could be devastated by even a moderate rise in sea level. These nations are densely populated, poor, and often cannot evacuate people in the event of a storm. Planning for sea level rise would not only save economic resources but human lives as well.

NOTES

1. All measurements in this chapter are presented in the metric system. Where doing so is not redundant, English equivalents are provided in parentheses. To avoid presenting a false sense of precision, this chapter translates entire idioms in several instances. Therefore, we translate "about a meter" into "a few feet" rather than into "about 3.3 feet."

2. Prehistoric shorelines have been found in the Mesabi Range in Minnesota (Sleep, 1976).

3. This term is technically a misnomer because a greenhouse prevents convectional, rather than radiational, cooling.

4. In a related effort, EPA held a symposium on the possible relationship between increased atmospheric levels of CO_2 and the frequency, severity, and track of hurricanes.

5. This calculation considered the lag between global temperature and thermal expansion of the ocean, but not the lag between temperature and ice melting (from computer printouts underlying Seidel and Keyes, 1983).

6. For example, the United States is expected to be responsible for only 14 percent of all CO_2 emissions by 2025, less than one decade's growth in emissions.

7. The model of the National Center for Atmospheric Research has recently estimated the warming to be nearly $4\,^\circ C$. See Warren M. Washington and Gerald A. Meehl, "Seasonal Cycle Experiments on the Climate Sensitivity Due to a Doubling of CO_2 with an Atmospheric General Circulation Model Coupled to a Simple Mixed Layer Ocean Model," NCAR/8041/82-1[E], Boulder: National Center for Atmospheric Research, August 1983, paper submitted to *Journal of Geophysical Research.*

8. This equation uses specified thermal sensitivity (the NAS estimates), greenhouse gas increases generated, and surface temperatures of the ocean and was developed by the Goddard Institute for Space Studies.

9. The conservative scenario assumed a diffusion coefficient of 1.18 cm^2/sec; the mid-range scenario, 1.54 cm^2/sec; and the high scenario, 1.9 cm^2/sec.

10. These special cases were run with coefficients of 0.2 cm^2/sec and 4.0 cm^2/sec.

11. The old assumptions that the case studies are based on, subsequently changed by Hoffman, are as follows: *Low:* Emissions of chlorofluorocarbons remain constant at the mid-1970s level; methane concentrations increase linearly by 2.0 percent of the 1980 level each year; and nitrous oxide concentrations increase linearly by 0.2 percent of the 1980 level each year. *High:* Concentrations of the trace gases all grow geometrically by 1.674 percent each year. These scenarios produced estimates of global sea level as follows: low—22.4 cm in 2025 and 74.6 cm in 2075; high—57.9 cm in 2025 and 219.3 cm in 2075.

12. As is generally the case with economic analyses conducted over a long period of time, the results are sensitive to the discount rate used to compute present values. If discount rates larger than the 3 percent assumed here are used, the economic impacts and value of anticipating sea level rise would be much smaller. Chapter 7 presents estimates using alternate discount rates and discusses the sensitivity of the results to various assumptions about investment behavior and community responses to sea level rise.

REFERENCES

Albanese, A., and M. Steinburg. 1980. *Environmental Control Technology for Atmospheric Carbon Dioxide.* Islip, N.Y.: Brookhaven National Laboratories.

Arrhenius, S. 1896. "On the Influence of Carbonic Acid in the Air upon the Temperature of the Ground." *Philus* 41:237.

Barnett, T. P. 1983. "Global Sea Level: Estimating and Explaining Apparent Changes." In Orville T. Magoon, ed., *Coastal Zone '83.* New York: American Society of Civil Engineers, pp. 2777-2795.

Barnett, T. P. (forthcoming). *Recent Changes in Sea Level and Their Possible*

Causes. LaJolla, Calif.: Climate Research Group, Scripps Institution of Oceanography.

Begley, Sharon et al., 1981. "Is Antarctica Shrinking?" *Newsweek.* October 5.

Bentley, Charles. 1980. "Response of the West Antarctic Ice Sheet to CO_2 Induced Climatic Warming." In *Environmental and Societal Consequences of a Possible CO_2-Induced Climate Change* (2) 1. Washington, D.C.: Department of Energy.

Bentley, C. R. 1983. "West Antarctic Ice Sheet: Diagnosis and Prognosis. " In *Proceedings: Carbon Dioxide Research Conference: Carbon Dioxide, Science and Consensus.* Conference 820970. Washington, D.C.: Department of Energy, pp. IV.3-IV.50.

Bloom, Arthur L. 1967. "Pleistocene Shorelines: A New Test of Isostasy." *Geological Society of America Bulletin* 78:1477-1494.

Boesch, D. F., ed. 1982. *Proceedings of the Conference on Coastal Erosion and Wetland Modification in Louisiana: Causes, Consequences and Options.* FWS-OBS-82/59. Washington, D.C.: Fish and Wildlife Service, Biological Services Program.

Broecker, W. S., Tsung-Hung Peng, and R. Engh. 1980. "Modeling the Carbon System." *Radiocarbon* 22:565-598.

Bruun, P. 1962. "Sea Level Rise as a Cause of Shore Erosion." *Journal of the Waterways and Harbors Division* 88(WW1):117-130.

Bruun, P. 1979. "The 'Bruun Rule,' Discussion on Boundary Conditions." In M. L. Schwartz and J. J. Fisher, eds., *Proceedings of the Per Bruun Symposium.* Newport, R.I.: IGU Commission on the Coastal Environment, pp.79-83.

Charney, J., chairman, Climate Research Board. 1979. *Carbon Dioxide and Climate: A Scientific Assessment.* Washington, D.C.: NAS Press.

Chylek, Peter, and William Kellogg. "The Sea Level Climate Connection." Unpublished paper. National Center for Atmospheric Research. Boulder, Co.

Clark, J. A., W. E. Farrell, and W. R. Peltier. 1978. "Global Changes in Postglacial Sea Level: A Numerical Calculation." *Quaternary Research* 9(3):265-287.

Colquhoun, D. J., M. J. Brooks, J. Michie, W. B. Abbott, F. W. Stapor, W. H. Newman, and R. R. Pardi. 1981. "Location of Archaeological Sites with Respect to Sea Level in the Southeastern United States." *Striae* 14:144-150.

Coolfont Conference. 1982. *1982 Department of Energy Research Conference.* Washington, D.C.: Department of Energy, Office of Carbon Dioxide Research.

Council on Environmental Quality. 1981. *Global Energy Futures and the Carbon Dioxide Problem.* Washington, D.C.: Government Printing Office.

Development Planning and Research Associates. 1982. Report to EPA/OSW.

Donn, W. L., W. R. Farrand, and M. Ewing. 1962. "Pleistocene Ice Volumes and Sea-Level Lowering." *Journal of Geology* 70:206-214.

Dubois, R. N. 1979. "Hypothetical Shore Profiles in Response to Rising Water Level." In M. L. Schwartz and J. J. Fisher, eds., *Proceedings of the Per Bruun Symposium.* Newport, R.I.:, IGU Commission on the Coastal Environment, pp. 13-31.

Dyson, F., 1976. *Can We Control the Amount of Carbon Dioxide in the Atmosphere?* Oak Ridge, Tenn.: Oak Ridge Associated Universities, Institute for Energy Analysis.

Edmonds, J., and J. Reilly. 1982. *Global Energy and CO$_2$ to the Year 2050.* Oak Ridge, Tenn.: Oak Ridge Associated Universities, Institute for Energy Analysis.

Emery, K. O. 1980. "Relative Sea Levels from Tide-Gage Records." *Proceedings, National Academy of Sciences, USA* 12:6968-6972.

Emery, K. O., and Robert H. Mead. 1971. "Sea Level as Affected by River Runoff, Eastern United States." *Science* 173:425-428.

Engineering-Science. 1979. *Evaluation of the Impacts of the Proposed Regulations to Implement the RCRA on Coal-Fired Electric Generating Facilities.* Prepared for the Department of Energy, contract number DE-AC 01-79ET135437, pp. XV and II-2.

Etkins, R., and E. Epstein. 1982. "The Rise of Global Mean Sea Level as an Indication of Climate Changes." *Science* 215:287-298.

Fairbanks, Richard G., and R. K. Matthews. 1978. "The Marine Oxygen Isotope Record in Pleistocene Coral, Barbados, West Indies." *Quaternary Research* 10(2):181-196.

Fairbridge, R. W., and W. S. Krebs, Jr. 1962. "Sea Level and the Southern Oscillation." *Geophysical Journal* 6:532-545.

Flohn, H. 1982. "Climate Change and an Ice-Free Arctic Ocean." In W. Clark, ed., *Carbon Dioxide Review: 1982.* New York: Oxford University Press, pp. 145-179.

Gornitz, V., S. Lebedeff, and J. Hansen. 1982. "Global Sea Level Trend in the Past Century." *Science* 215:1611-1614.

Greenberg, Don. 1982. "Sequestering," Unpublished memorandum prepared for Office of Policy Analysis, Environmental Protection Agency, Washington, D.C.

Gutenberg, Beno. 1941. "Changes in Sea Level, Postglacial Uplift, and Mobility of the Earth's Interior." *Bulletin of the Geological Society of America* 52:721-772.

Hands, E. B. 1979. "Bruun's Concept Applied to the Great Lakes." In M. L. Schwartz and J. J. Fisher, eds., *Proceedings of the Per Bruun Symposium.* Newport, R.I.: IGU Commission on the Coastal Environment, pp.63-66.

Hands, E. B. 1981. *Predicting Adjustments in Shore and Offshore Sand Profiles on the Great Lakes,* CERC technical aid 81-4, Fort Belvoir, Va.: Coastal Engineering Research Center.

Hansen, J. 1982. *Global Climate Change Due to Increasing CO$_2$ and Trace Gases.* Washington, D.C.: Statement to the U.S. House of Representatives Subcommittee on Natural Resources, Agriculture Research and Environment, and the Subcommittee on Investigations and Oversight of the Science and Technology Committee.

Hansen, J., D. Johnson, A. Lacis, S. Lebedeff, D. Rind, and G. Russell. 1981. "Climate Impact of Increasing Atmospheric Carbon Dioxide." *Science* 213:957-966.

Harmon, Russel S., Lynton S. Land, Richard M. Mitterer, Peter Garrett, Henry P. Schwarcz, and Grahame J. Larson. 1981. "Bermuda Sea Level during the Last Interglacial." *Nature* 289:481-483.

Harmon, Russel S., Henry P. Schwarcz, and Derek C. Ford. 1978. "Late Pleistocene Sea Level History of Bermuda." *Quaternary Research* 9(2):205-218.

Haskin, H. H., and S. M. Tweed. 1976. *Oyster Setting and Early Spat Survival at Critical Salinity Levels on Natural Seed Oyster Beds of Delaware Bay.* New Brunswick, N.J. Water Resources Research Institute, Rutgers University.

Hays, J. D., and W. C. Pitman III. 1973. "Lithospheric Plate Motion, Sea-Level Changes, and Climatic and Ecological Consequences." *Nature* **246**:18-22.

Hey, R. W. 1978. "Horizontal Quaternary Shorelines of the Mediterranean." *Quaternary Research* **10**:197-203.

Hicks, Steacy D. 1978. "An Average Geopotential Sea Level Series for the United States." *Journal of Geophysical Research* **83**(C3):1377-1379.

Hicks, Steacy D. 1981. "Long-Period Sea Level Variations for the United States through 1978." *Shore and Beach* **26**:9.

Hoffman, J., D. Keyes, and J. Titus. 1983. *Projecting Future Sea Level Rise: Methodology, Estimates to the Year 2100, and Research Needs.* 2nd rev. ed. U.S. GPO No. 055-000-0236-3. Government Printing Office.

Hollin, John T. 1983. "Climate and Sea Level in Isotope Stage 5: An East Antarctic Ice Surge at 95,000 BP?" *Nature* **283**:629-633.

Hollin, John T., and R. Barry. 1979. "Empirical and Theoretical Evidence Concerning the Response of the Earth's Ice and Snow Cover to a Global Temperature Increase." *Environmental International* **2**:437-444.

Hopkins, David M. 1973. "Sea Level History in Beringia during the Past 250,000 Years." *Quaternary Research* **3**:520-540.

Hughes, T. 1983. "The Stability of the West Antarctic Ice Sheet: What Has Happened and What Will Happen." In *Proceedings: Carbon Dioxide Research Conference: Carbon Dioxide, Science and Consensus, 1983.* DOE Conference 820970. Washington, D.C.: Department of Energy, pp.IV.51-IV.73.

Hughes, T., J. L. Fastook, and G. H. Denton. 1979. *Climatic Warming and the Collapse of the West Antarctic Ice Sheet.* Orono, Maine: University of Maine.

Hull, C. H. J., and R. Tortoriello. 1979. "Sea-Level Trend and Salinity in the Delaware Estuary." Staff paper. West Trenton, N.J. : Delaware River Basin Commission.

Institute for Energy Analysis. 1981. *Determinants of Global Energy Supply to the Year 2050.* Oak Ridge, Tenn.: Oak Ridge Associated Universities.

Keeling, C. D., R. B. Bacastow, and T. P. Whorf. 1982. "Measurements of the Concentration of Carbon Dioxide at Mauna Loa Observatory, Hawaii." In W. Clark, ed., *Carbon Dioxide Review: 1982.* New York: Oxford University Press, pp.377-384.

Keu, Richard A. 1981. "Whither the Shoreline." *Science* **214**:428.

Keyfitz, N., E. Allen, J. Edmonds, R. Dougher, and B. Widget. 1983. *Global Population, 1975-2075, and Labor Force, 1975-2050.* Oak Ridge, Tenn.: Oak Ridge Associated Universities, Institute for Energy Analysis.

Kopec, Richard J. 1971. "Global Climate Change and the Impact of a Maximum Sea Level on Coastal Settlement." *Journal of Geography* **70**:541-550.

Ku, Teh-Lung, Margaret A. Kimmel, William H. Easton, and Thomas J. O'Neil. 1974. "Eustatic Sea Level 120,000 Years Ago on Oahu, Hawaii." *Science* **183**:959-962.

Kukla, G., and J. Gavin. 1981. "Summer Ice and Carbon Dioxide." *Science* 214:497-503.

LaBrecque, John L., and Peter Barker. 1981. "The Age of the Weddell Basin." *Nature* 290:489-492.

Lacis, A., J. Hanson, P. Lee, T. Mitchell, and S. Lebedeff. 1981. "Greenhouse Effect of Trace Gases, 1970-1980." *Geophysical Research Letters* 8(10):1035-1038.

Lamb, H. H. 1970. "Volcanic Dust in the Atmosphere." *Philosophical Transactions of the Royal Society, London,* series A, 255:425-533.

Leatherman, Stephen P., Michael S. Kearny, and Beach Clow. 1983. *Assessment of Coastal Responses to Projected Sea Level Rise: Galveston Island and Bay, Texas.* URF report TR-8301; report to ICF under contract to EPA. College Park, Md.: University of Maryland.

Lisitzin, Eugenie. 1974. *Sea-Level Changes,* Elsevier Oceanography Series 8. New York: Elsevier Scientific Publishing Co.

Lugo, Ariel. 1980. "Are Tropical Forest Ecosystems a New Sink?" In *The Role of Tropical Forests in the World Carbon Cycle.* Washington, D.C.: Department of Energy, pp.1-18.

Manabe, S., and R. J. Stouffer. 1980. "Sensitivity of a Global Climate Model to an Increase of CO_2 Concentration in the Atmosphere." *Journal of Geophysical Research* 85:5529-5554.

Marshall, J. F., and B. G. Thom. 1976. "The Sea Level in the Last Interglacial." *Nature* 263:120-121.

Massachusetts, State of. 1981. *Barrier Beaches: A Few Questions Answered.* Boston: Massachusetts Office of Coastal Zone Management.

Mercer, J. H. 1970. "Antarctic Ice and Interglacial High Sea Levels." *Science* 168:1605-1606.

Mercer, J. H. 1972. "The Lower Boundary of the Holocene." *Quaternary Research* 2(1):15-24.

Mercer, J. H. 1976. "Glacial History of Southernmost South America." *Quaternary Research* 6:125-166.

Mercer, J. H. 1978. "West Antarctic Ice Sheet and CO_2 Greenhouse Effect: A Threat of Disaster." *Nature* 271(5643):321-325.

Nardin, Thomas R., Robert H. Osborne, David J. Bottjer, and Robert C. Scheidemann. 1981. "Holocene Sea Level Curves and Santa Monica Shelf, California Continental Borderland." *Science* 213:331-333.

New Jersey, Department of Environmental Protection. 1981. *New Jersey Shore Protection Master Plan.* Trenton: Division of Coastal Resources.

Oerlemans, J. 1981. "Effect of Irregular Fluctuations in Antarctic Precipitation on Global Sea Level." *Nature* 290:770-772.

Palmer, A., W. Moose, T. Quinn, K. Wolf. 1980. *Economic Implications for Regulating Chlorofluorocarbons Emissions for Non-Aerosol Applications.* R-2524-EPA. Washington, D.C.: Environmental Protection Agency.

Perry, A. M., K. J. Araj, W. Fulkerson, D. J. Rose, M. M. Miller, and R. M. Rotty. 1981. *Energy Supply and Demand Implications of CO_2.* Oak Ridge, Tenn: Oak Ridge National Laboratories.

Pilkey, O., J. Howard, B. Brenninkmeyer, R. Frey, A. Hine, J. Kraft, R. Morton, D. Nummedal, and H. Wanless. 1981. *Saving the American Beach: A Position*

Paper by Concerned Coastal Geologists. Results of the Skidaway Institute of Oceanography Conference on America's Eroding Shoreline. Savannah: Skidaway Institute of Oceanography.

Revelle, R., 1983. "Probable Future Changes in Sea Level Resulting from Increased Atmospheric Carbon Dioxide," in *Changing Climate.* Carbon Dioxide Assessment Committee. Washington, D.C.: National Academy Press, pp.433-447.

Research Planning Institute. 1983. *Hypothetical Shoreline Changes Associated with Various Sea-Level Rise Scenarios for the United States: Case Study— Charleston, South Carolina.* Report to ICF under contract to EPA. Columbia, S. C.: RPI.

Rotty, R., and G. Marland. 1980. *Constraints on Carbon Dioxide Production From Fossil Fuel Use.* ORAM-IEA-80-9(m). Oak Ridge, Tenn.: Oak Ridge Associated Universities, Institute for Energy Analysis.

Ruch, Carlton. 1982. *Ongoing Research on the Effects of Hurricanes.* College Station, Tex.: Research Center, College of Architecture, Texas A&M University.

Schneider, Stephen H., and Robert S. Chen. 1980. "Carbon Dioxide Flooding: Physical Factors and Climatic Impact." *Annual Review of Energy* 5:107-140.

Schwartz, M. L. and J. J. Fisher, eds. 1979. *Proceedings of the Per Bruun Symposium.* Newport, R.I.: IGU Commission on the Coastal Environments.

Seidel, S., and D. Keyes. 1983. *Can We Delay a Greenhouse Warming?* Washington, D.C.: Government Printing Office.

Sleep, N. H. 1976. "Platform Subsidence Mechanisms and 'Eustatic' Sea-Level Change." *Techtonophysics* 36(1-3):45-56.

Smagorinsky, J., chairman, Climate Research Board. 1982. *Carbon Dioxide: A Second Assessment.* Washington, D.C.: NAS Press.

Sorensen, Robert M., Richard N. Weisman, and Gerard P. Lennon. 1983. *Methods for Controlling the Increases in Shore Erosion Inundation by Storm Surge and Salinity Intrusion Caused by a Postulated Sea-Level Rise.* Report to ICF under contract to EPA. Bethlehem, Pa.:Lehigh University, Department of Civil Engineering.

Sullivan, Walter. 1981. "Study Finds Warming Trend That Could Raise Sea Levels." *New York Times,* August 22, p. 1, col. 1.

Thomas, R., T. J. O. Sanderson, and K. E. Rose. 1979. "Effect of Climatic Warming on the West Antarctic Ice Sheet." *Nature* 277:355-362.

U.S. Department of the Interior. 1983. *Undeveloped Coastal Barriers: Final Environmental Statement.* Washington, D.C.: Coastal Barriers Task Force.

U.S. Fish and Wildlife Service. 1979. *The Effect of Salinity Change on the American Oyster in Delaware Bay.* Prepared for the Army Corps of Engineers, Philadelphia District. Delaware Salinity Intrusion Study, Fish and Wildlife Service, Department of Interior, State College, Pa.

Worsley, Thomas R., and Thomas A. Davies. 1979. "Sea Level Fluctuations and Deep-Sea Sedimentation Rates." *Science* 203:455-456.

APPENDIX

Summaries of Charleston, South Carolina and Galveston, Texas, Case Studies

Table1-A. Charleston Study Area: Area Lost to Sea Level Rise by Scenario (in sq. km)

	1980	2025				2075			
	Area	Trend	Low	Medium	High	Trend	Low	Medium	High
Charleston Peninsula	27.4	0.5	1.0	1.8	2.8	1.0	2.8	7.5	13.2
Mount Pleasant	29.8	0.5	1.3	2.1	3.6	0.8	3.6	6.2	10.4
Sullivans Island	2.8	_[a]	0.3	0.5	0.8	_[a]	1.0	2.1	2.3
West Ashley/ James Island	14.0	0.5	1.0	1.8	3.1	1.0	4.9	6.0	9.3
Daniel Island/ Naval Base/ Marsh	19.7	_[a]	1.3	1.6	2.6	0.8	2.8	7.0	7.8
Total	93.7	1.8	4.9	7.8	13.0	3.9	14.2	28.7	43.0

[a]Less than 0.1 sq km.

Table 1-B. Charleston Study Area: Area in 10-Year and 100-Year Floodplains by Scenario (in sq. km)

	Total Area (1980)	1980	2025				2075			
			Trend	Low	Medium	High	Trend	Low	Medium	High
10-Year Floodplain										
Charleston Peninsula	27.4	5.4	6.0	7.0	7.8	8.5	6.7	9.8	14.8	18.9
Mount Pleasant	29.8	8.3	8.8	9.3	9.8	10.4	9.3	11.1	14.2	16.8
Sullivans Island	2.8	2.3	2.3	2.6	2.6	2.6	2.6	2.8	2.8	2.8
West Ashley/ James Island	14.0	6.5	6.7	7.2	7.5	7.8	7.0	8.5	10.9	12.7

Table 1-B. *(continued)*

	Total Area (1980)	1980	2025 Trend	Low	Medium	High	2075 Trend	Low	Medium	High
			10-Year Floodplain (continued)							
Daniel Island/ Naval Base/ Marsh	19.7	8.3	8.8	9.8	10.6	11.6	9.3	12.6	16	18.1
Total	93.7	30.8	32.9	35.7	38.6	41.4	34.9	45.0	58.5	69.4
			100-Year Floodplain							
Charleston Peninsula	27.4	14.7	15.3	16.3	17.3	18.1	16.0	19.4	22.3	24.0
Mount Pleasant	29.8	14.2	14.8	15.5	16.1	16.8	15.3	17.6	20.7	23.8
Sullivans Island	2.8	2.8	2.8	2.8	2.8	2.8	2.8	2.8	2.8	2.8
West Ashley/ James Island	14.0	11.4	11.7	12.2	12.4	12.7	11.9	12.9	13.7	14.0
Daniel Island/ Naval Base/ Marsh	19.7	16.1	16.6	17.1	17.6	17.9	16.8	18.4	19.2	19.4
Total	93.7	59.2	61.1	63.7	66.0	68.4	62.9	71.2	78.7	83.9

Table 1-C. Potential Storm Damage and Inundation Losses Under Various Sea Level Rise Scenarios For the Charleston Study Area (in millions of 1980$)

Damage	1980	2025 Trend	Low	Medium	High	2075 Trend	Low	Medium	High
Potential damage from the 100 year storm	316	510	555	600	640	620	800	800	720[b]
Expected annual damage across all storms[a]	13	23	25	39	45	32	62	67	64

(continued)

[a]Expected value equals the sum of the damage for each storm times the probability of each storm.

[b]Storm surge damage under 2075 high scenario is lower than that for low and medium scenarios because so much area would already be lost to shoreline retreat.

Table 1-C. *(continued)*

Damage	1980	2025 Trend	Low	Medium	High	2075 Trend	Low	Medium	High
Total value of land and structures lost by shoreline retreat for all years, 1980–2075	-	1	7	11	35	6	60	420	870

Note: All estimates are made under the assumptions that development proceeds at rates currently anticipated in the absence of sea level rise and that no additional protective structures are built.

Table 1-D. Galveston Study Area: Area Lost to Sea Level Rise by Scenario (in km, sq mi in parentheses)

City	1980 Area	2025 Trend	Low	Medium	High	2075 Trend	Low	Medium	High
San Leon	19.4 (7.5)	0.3 (0.1)	0.5 (0.2)	1.0 (0.4)	1.3 (0.5)	1.3 (0.5)	3.4 (1.3)	5.2 (2.0)	7.0 (2.7)
Galveston Island	69.2 (26.7)	0.5 (0.2)	1.3 (0.5)	2.1 (0.8)	2.6 (1.0)	1.8 (0.7)	3.9 (1.5)	6.5 (2.5)	8.3 (3.2)
Texas City, La Marque, other	186.5 (72.0)	1.8 (0.7)	2.3 (0.9)	3.4 (1.3)	4.4 (1.7)	3.1 (1.2)	8.3 (3.2)	12.7 (4.9)	17.1 (6.6)
Total	275.1 (106.2)	2.6 (1.0)	4.1 (1.6)	6.5 (2.5)	8.3 (3.2)	6.2 (2.4)	15.6 (6.0)	24.4 (9.4)	32.4 (12.5)

Table 1-E. Galveston Study Area in 15-Year and 100-Year Floodplains by Scenario (in sq km, sq mi in parentheses)

City	Total Area (1980)	1980	2025				2075			
			Trend	Low	Medium	High	Trend	Low	Medium	High
15-Year Floodplain										
San Leon	19.4 (7.5)	1.0 (0.4)	1.6 (0.6)	1.8 (0.7)	2.1 (0.8)	2.1 (0.8)	1.8 (0.7)	2.3 (0.9)	4.4 (1.7)	5.4 (2.1)
Galveston Island	69.2 (26.7)	18.9 (7.3)	21.2 (8.2)	24.2 (9.3)	26.4 (10.2)	28.5 (11.0)	23.8 (9.2)	31.1 (12.0)	44.5 (17.2)	58.8 (22.7)
Texas City, La Marque, other	186.5 (72.0)	45.3 (17.5)	48.9 (18.9)	52.3 (20.2)	54.4 (21.0)	55.9 (21.6)	51.8 (20.0)	58.5 (22.6)	70.7 (27.3)	78.5 (30.3)
Total	275.0 (106.2)	65.3 (25.2)	71.7 (27.7)	78.2 (30.2)	82.9 (32.0)	86.5 (33.4)	77.4 (29.9)	91.9 (35.5)	119.7 (46.2)	142.7 (55.1)
100-Year Floodplain										
San Leon	19.4 (7.5)	16.1 (6.2)	17.4 (6.7)	18.1 (7.0)	18.9 (7.3)	19.2 (7.4)	18.1 (7.0)	19.4 (7.5)	19.4 (7.5)	19.4 (7.5)
Galveston Island	69.2 (26.7)	45.3 (17.5)	45.8 (17.7)	46.6 (18.0)	47.1 (18.2)	50.8 (19.6)	46.4 (17.9)	60.1 (23.2)	62.9 (24.3)	62.3 (24.4)
Texas City, La Marque, other	186.5 (72.0)	99.2 (38.3)	100.0 (38.6)	100.7 (38.9)	101.0 (39.0)	137.3 (53.0)	100.7 (38.9)	179.2 (69.2)	184.9 (71.4)	186.5 (72.0)
Total	275.0 (106.2)	160.6 (62.0)	163.2 (63.0)	165.5 (63.9)	167.0 (64.5)	206.9 (79.9)	165.2 (63.8)	258.7 (99.9)	267.3 (103.2)	269.1 (103.9)

Table 1-F. Potential Storm Damage and Inundation Losses under Various Sea Level Rise Scenarios for the Galveston Study Area (in millions of 1980$)

Damage	1980	2025				2075			
		Trend	Low	Medium	High	Trend	Low	Medium	High
Potential damage from the 100-year storm	260	535	580	600	1,100	600	1,800	2,100	2,400
Expected annual damage across all storms[a]	6	16	22	27	38	23	57	105	170
Total value of land and structures lost by shoreline retreat for all years 1980-2075	-	2	6	12	17	17	49	87	107

Note: All estimates were made under the assumptions that development proceeds at rates currently anticipated in the absence of sea level rise and that no additional protective structures are built.

[a]Expected value equals the sum of the damage for each storm times the probability of each storm.

Chapter 2

Climate Sensitivity to Increasing Greenhouse Gases

James E. Hansen, Andrew A. Lacis,
David H. Rind, and Gary L. Russell

INTRODUCTION

Climate changes occur on all time scales, as illustrated in Figure 2-1 by the trend of global mean surface air temperature in the past century, the past millennium, and the past 30,000 years. The range of global mean temperature in the past 30,000 years and indeed the past million years has been of the order of 5°C. At the peak of the last glacial period, the Wisconsin ice age approximately 18,000 years ago, the mean temperature was 3-5°C (5-9°F) cooler than today. At the peak of the current interglacial, 5,000-8,000 years ago, the mean temperature is estimated to have been 0.5-1°C warmer than today (Figure 2-1). In the previous (Eemian) interglacial, when sea level is thought to have been about 5m higher than today (Hollin, 1972), global mean temperature appears to have been of the order of 1°C warmer than today.

Global mean temperature is a convenient parameter, but it must be recognized that much larger changes may occur on more localized scales. Decadel variations of global temperature in the past century, for example, are enhanced by about a factor of three at high latitudes (Hansen et al., 1983a). Also, the global cooling of 3-5°C (5-9°F) during the Wisconsin ice age included much larger regional changes, as evidenced by the ice

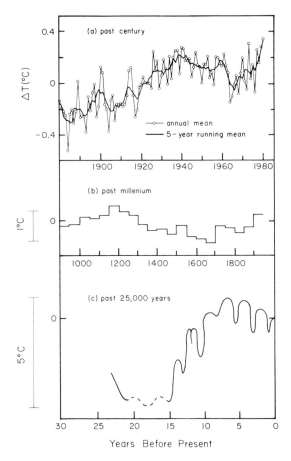

Figure 2-1. Global temperature trend for (a) past century, (b) millennium, and (c) 25,000 years. (a) is based on J. Hansen, D. Johnson, A. Lacis, S. Lebedeff, P. Lee, D. Rind, and G. Russell, 1981, "Climate Impact of Increasing Atmospheric Carbon Dioxide," *Science* **213:**957–966, updated through 1981. (b) is based on temperatures in central England, the tree limit in the White Mountains of California, and oxygen isotope measurements in the Greenland ice (W. Dansgaard of the Geophysical Isotope Laboratory, University of Copenhagen, pers. comm.), with the temperature scale set by the variations in the last 100 years. (c) is based on changes in tree lines, fluctuations of alpine and continental glaciers, and shifts in vegetation patterns recorded in pollen spectra (National Academy of Sciences, 1975. *Understanding Climatic Change,* Washington, D.C.: National Academy Press), with the temperature scale set by the 3–5° cooling obtained in a 3-D climate model J.E. Hansen et al., 1983b, "Efficient Three Dimensional Global Models for Climate Studies: Models I and II," *Monthly Weather Review* **111:**609–662; J. Hansen et al., 1984; "Climate Sensitivity: Analysis of Feedback Mechanisms," in *Climate Processes and Climate Sensitivity,* J.E. Hansen and T. Takahashi, eds., Washington, D.C.: American Geophysical Union, pp. 130–163 with the boundary conditions for 18,000 years ago. Thus, the shapes of curves (b) and (c) are based on only Northern Hemisphere data.

sheet of 2 km (1.3 mi) mean thickness covering much of North America including the present sites of New York, Minneapolis, and Seattle.

The recorded climate variations include the response to external forcings (e.g., changes in the amount or global distribution of solar irradiance) and also internal climate fluctuations (e.g., changes in ocean dynamics driven by weather "noise"). Determination of the division of actual climate variations between these two categories is a fundamental task of climate investigations.

The mean temperature of the earth is determined primarily by the amount of energy absorbed from the sun, which must be balanced on average by thermal emission. The earth's surface temperature also depends on the atmosphere, which partially blankets the thermal radiation and thus requires the surface to be hotter in order for the thermal emission to balance the absorbed solar radiation. Today the mean temperature of the earth's surface is 288K, 33°C higher than it would be in the absence of this "greenhouse" blanketing by the atmosphere.

As the CO_2 content of the atmosphere increases, the atmosphere becomes more opaque at infrared wavelengths where CO_2 has absorption bands, thus raising the mean level of emission to space to higher altitudes. A simple radiative calculation shows that doubling atmospheric CO_2 would raise the mean level of emission to space, averaged over the thermal emission spectrum, by about 200m. (Cf. discussion in the section below on empirical evidence of climate sensitivity.) Since atmospheric temperature falls off with altitude by about 6°C/km, the planet would have to warm by about 1.2°C to restore equilibrium if the tropospheric temperature gradient and other factors remained unchanged. In general, other factors would not remain unchanged, and thus the actual temperature change at equilibrium would differ from the one in this simple calculation by some "feedback" factor, f,

$$\Delta T_{eq} = f\, \Delta T_{rad} \tag{2.1}$$

where ΔT_{eq} is the equilibrium change in global mean surface air temperature and ΔT_{rad} is the change in surface temperature that would be required to restore radiative equilibrium if no feedbacks occurred.

The feedback factor f not only determines the magnitude of the eventual climate change for a given change in climate forcing but also the time required to approach the new equilibrium. The reason for this is the fact that the initial rate at which the ocean warms is determined by only the magnitude of the direct climate forcing, that is, the feedbacks only come into play as the warming occurs, and thus the ocean thermal response time increases with increasing f (Hansen et al., 1981, 1984). The physical processes expected to contribute to the feedback factor include the ability of the atmosphere to hold more water vapor (which is also a greenhouse gas) with increasing temperature and the change of snow and ice cover (and thus albedo) with changing temperature.

In this chapter we first discuss current climate model evidence for climate sensitivity, which suggests a range of $3\pm1.5°C$ for doubled CO_2, corresponding to a net feedback factor $f \sim 2.5$. We then summarize empirical evidence for climate sensitivity and feedback processes, which provide substantial support for the magnitude of climate effects computed by the models. Finally, we look at current trends of greenhouse gases and global temperature, which allow us to discuss the magnitude of warming expected in coming decades.

CLIMATE MODEL CALCULATIONS OF CLIMATE SENSITIVITY

Two National Academy of Sciences panels (Charney, 1979; Smagorinsky, 1982) estimated the equilibrium (global mean) climate sensitivity for doubled CO_2 to be $3\pm1.5°C$. This conclusion was based on consideration of the primary mechanisms believed to contribute to global climate sensitivity, including the study of results from two 3-D global climate models: that of Manabe and Stouffer (1980), which yields a sensitivity near $2°C$, and that of Hansen et al. (1983b, 1984), which yields a sensitivity near $4°C$.

In this section we illustrate the temperature change produced in the latter 3-D climate model when CO_2 is doubled and analyze the physical mechanisms contributing to this sensitivity. This provides a basis for discussing the uncertainties in the computed climatic sensitivity due to approximations in representing these processes. The temperature change computed for doubled CO_2 is shown in Figure 2-2 in three ways: (a) the annual mean surface air temperature as a function of latitude and longitude; (b) the zonal mean surface air temperature as a function of latitude and month; and (c) the annual and zonal mean temperature as a function of altitude and latitude.

The surface air warming is enhanced at high latitudes (Figure 2-2a) partly because of confinement of the greenhouse warming to lower layers as a consequence of the atmospheric stability at high latitudes and partly because of the ice/snow albedo feedback at high latitudes. The enhanced warming in the African and Australian deserts can be traced to the stability of the atmosphere above these regions and the relative lack of evaporative cooling. The maximum warming near West Antartica is associated with the largest reduction in sea ice cover there. Many aspects of the geographical distribution of the warming for doubled CO_2 are

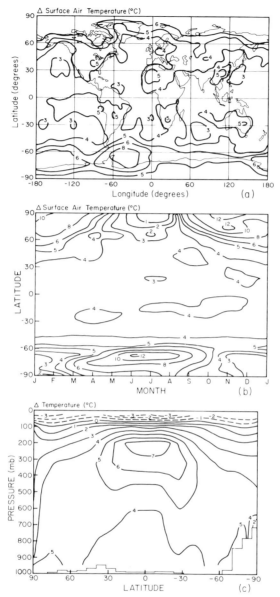

Figure 2-2. Warming of air temperature due to doubled CO_2 in the 3-D global climate model of Hansen et al. (a) shows the geographical distribution of annual mean surface air warming; (b) shows the seasonal variation of the surface air warming averaged over longitude; and (c) shows the altitude distribution of the warming averaged over season and longitude. (After J. Hansen, A. Lacis, D. Rind, G. Russell, P. Stone, I. Fung, R. Ruedy, and J. Lerner, 1984. "Climate Sensitivity: Analysis of Feedback Mechanisms." in *Climate Processes and Climate Sensitivity,* J. E. Hansen and T. Takahashi, eds., Washington, D.C.: American Geophysical Union, pp. 130–163.)

clearly related to changes in prevailing wind patterns. However, the detailed geographical patterns of the computed climate changes should not be viewed as a reliable prediction for a doubled CO_2 world, because current climate models still poorly represent many parts of the climate system. For example, changes in horizontal heat transport by the oceans, which will undoubtedly influence regional climate patterns, are not included in the simulations for doubled CO_2.

The strong seasonal variation of the computed warming at high latitudes (Figure 2-2*b*) is due to the seasonal change of atmospheric stability and the influence of melting sea ice in the summer, which limits the ocean temperature rise. At low latitudes, the temperature rise is greatest in the

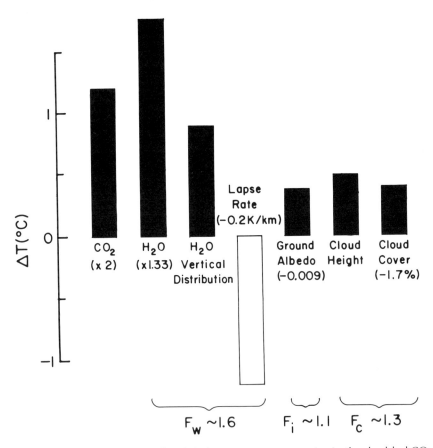

Figure 2-3. Contributions to the global mean temperature rise in the doubled CO_2 experiment as estimated by inserting changes obtained in the 3-D experiment into a 1-D radiative/convective climate model.

upper troposphere (Figure 2-2c), because the added heating at the surface primarily causes increased evaporation and moist convection, with resultant deposition of latent heat and water vapor at high levels. These processes are discussed further below.

The processes in this 3-D climate model that are responsible for the 4°C temperature rise for doubled CO_2 can be investigated with the help of a 1-D model of Lacis et al. (1981), inserting into it one-by-one all the radiatively significant global mean changes that were observed to occur in the 3-D experiments. This analysis procedure and its limitations are discussed in greater detail elsewhere (Hansen et al., 1984). The contributions found for the different changes that occurred in the 3-D model are illustrated in Figure 2-3.

Water vapor increase provides the dominant feedback, with most of the effect given by the increase in mean water vapor amount. Additional positive feedback occurs because the water vapor distribution is weighted more to higher altitudes for the doubled CO_2 case. However, the change in lapse rate, mainly due to the added H_2O, almost cancels the effect of the change in the water vapor vertical profile. Since the amount of water the atmosphere holds is largely dependent on the mean temperature, it is expected that the latter two effects would approximately cancel. Thus, it seems unlikely that the net water vapor feedback factor can be greatly in error, even though the water vapor distribution and lapse rate depend on the moist convection process, which is difficult to model realistically.

Ground albedo decrease also provides a substantial feedback (Figure 2-3). The ground albedo change (Figure 2-4a) is largely due to reduced sea ice. Shielding of the ground by clouds and the atmosphere (Figure 2-4b) makes this feedback several times smaller than it would be in the absence of the atmosphere. However, it is a significant positive feedback, and for this model it is at least as large in the Southern Hemisphere as in the Northern Hemisphere. The 1-D RC model does not provide a complete analysis of the sea ice feedback, for example, of the effect of sea ice in insulating the ocean and thus reducing radiation to space. However, from the geographic pattern of the temperature increase (Figure 2-2), and the coincidence of warming maxima with reduced sea ice, it is clear that the sea ice provides a positive feedback.

Cloud changes (Figure 2-5) also provide a significant positive feedback for doubled CO_2 in this model, as a result of a small increase (about 10mb) in mean cloud height and a 1.7 percent decrease in cloud cover. Present understanding of cloud processes does not permit confirmation or contradiction of the realism of these changes. The increase of mean cloud height is plausible: it falls between the common assumptions of fixed cloud height and fixed cloud temperature. The change in cloud cover reflects a reduction of low and middle level clouds, associated with a drying of

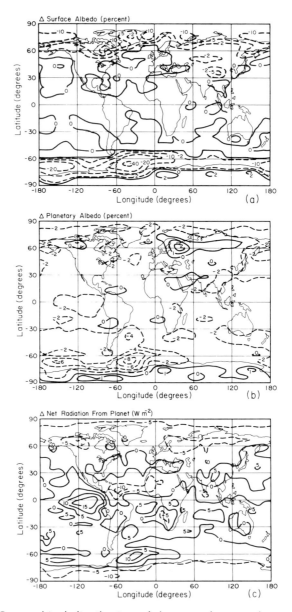

Figure 2-4. Geographical distribution of the annual mean changes of surface albedo, planetary albedo, and net radiation from the planet due to doubled CO_2 in the 3-D global climate model of Hansen et al. (After J. Hansen, A. Lacis, D. Rind, G. Russell, P. Stone, I. Fung, R. Ruedy, and J. Lerner, 1984. "Climate Sensitivity: Analysis of Feedback Mechanisms." in *Climate Processes and Climate Sensitivity,* J.E. Hansen and T. Takahaski, eds., Washington, D. C.: American Geophysical Union, pp. 130–163.)

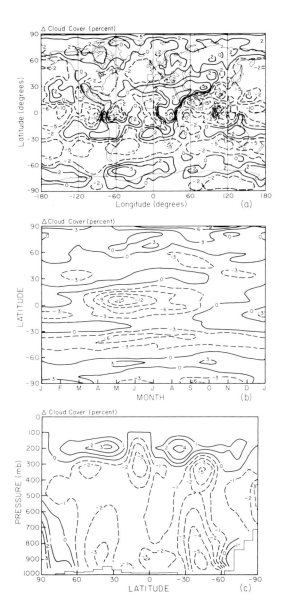

Figure 2-5. Cloud cover changes due to doubled CO_2 in the 3-D global climate model of Hansen et al. (a) show the geographical distribution of annual mean cloud cover changes; (b) shows the seasonal variation of cloud cover changes averaged over longitude; and (c) shows the altitude distribution of cloud cover changes, averaged over season and longitude. (After J. Hansen, A. Lacis, D. Rind, G. Russell, P. Stone, I. Fung, R. Ruedy, and J. Lerner, 1984. "Climate Sensitivity: Analysis of Feedback Mechanisms." in *Climate Processes and Climate Sensitivity*, J. E. Hansen and T. Takahashi, eds., Washington, D. C.: American Geophysical Union, pp. 130–163.)

these layers due to an increase of penetrating cumulus convection. Improved assessment of the cloud contribution will depend primarily on the development of increasingly realistic representations of cloud formation processes in global climate models, as verified by an accurate global cloud climatology.

The processes providing the major feedbacks in this climate model are thus atmospheric water vapor, clouds, and the surface albedo. Considering the earth from a planetary perspective, it seems likely that these are the principal feedback processes for the earth on a time scale of decades. The albedo of the planet for solar radiation is primarily determined by the clouds and the surface, with the main variable component of the latter being the ice/snow cover. The thermal emission of the planet is primarily determined by the atmospheric water vapor and clouds. Thus, those processes principally responsible for the earth's radiation balance and temperature are included in the model and are responsible for the significant feedbacks in the model.

There is substantial uncertainty in the quantitative value of these feedbacks. However, the most important feedback, due to water vapor, seems certain to be greater than one and is unlikely to be less than approximately 1.5. The ice/snow albedo feedback seems certain to be greater than one. The cloud feedback could be greater or less than one. Our model suggests that it is a significant positive ($f > 1$) feedback, but much more work is needed.

These feedback factors suggest some sources for the difference between our climate model sensitivity and that of Manabe and Stouffer (1980). They use fixed clouds (altitude and cloud cover) and thus have $f_{cloud} \equiv 1$. Also, their control run has less sea ice than our model, so that their feedback factor for that process should be between one and the value for our model. Therefore, it is likely that their primary feedback is $f_{water\ vapor}$, and it is not surprising that their sensitivity is approximately $2°$ C for doubled CO_2.

Although the cloud and sea ice feedbacks appear to "account" for most of the difference in sensitivity between our model and that of Manabe and Stouffer, we point out that there is another major difference between the models. This difference relates to the atmosphere and ocean transports of energy, whose feedbacks do not show up as identified components in an energy balance analysis such as in Figure 2-3. Our model includes a specified horizontal transport of heat by the ocean, which is identical in the control and experiment runs; thus there is no ocean tranport feedback in our model.

Manabe and Stouffer do not explicitly allow feedback on ocean transport either, because the ocean transport is zero in both experiment and control runs. However, in their model, increased poleward transport of energy in the atmosphere apparently replaces poleward transport of heat in the ocean, since their high latitude regions are at least as warm as in our model (and observations). This surrogate oceanic transport (in the atmosphere) may provide a negative feedback; the decrease in latitudinal temperature gradient accompanying a warmer atmosphere generally tends to decrease atmospheric transports, thus providing a negative feedback (Stone, 1984). Thus, while our ocean transport has no feedback effect, being identical in experiment and control runs, Manabe and Stouffer's surrogate transport probably has a negative feedback; indeed, Manabe and Wetherald (1975, 1980) explicitly show a negative feedback poleward of mid-latitudes for doubled CO_2 runs with idealized topography. The contribution of this feedback could be quantified by running the same model with and without fixed ocean transport.

In summary, available global climate models all suggest an equilibrium global climate sensitivity in the range of 2-4°C for doubled CO_2. This range is consistent with that estimated by the National Academy of Sciences, 1.5-4.5°C, which attempted to allow for uncertainties not accounted for in existing models. It is certainly conceivable that the true climate sensitivity is outside this range. However, a sensitivity smaller than 1.2°C would require the hypothesis of a net negative feedback. Such a hypothesis, though it can not be ruled out a priori, is faced with the difficult task of finding a negative feedback strong enough to overcome the dominant feedback mechanism that has been identified, that is, the ability of the atmosphere to hold more water vapor at higher temperatures, which is strongly positive ($f_{water\ vapor} \sim 1.6$).

Improvement of the ability of global climate models to realistically simulate climate change will require better understanding of key physical processes such as moist convection, large-scale cloud formation and ocean circulation, including its response to a warming of the ocean mixed layer. Better understanding of these processes, in turn, depends on appropriate observations from both global-scale and small-scale studies.

EMPIRICAL EVIDENCE OF CLIMATE SENSITIVITY

Planetary Data

A valuable test of the magnitude of the greenhouse effect can be obtained by examining the ensemble of experiments provided by the conditions on several different planets. The terrestrial planets Mars, Earth, and Venus provide a particularly appropriate set because they have a broad range of abundances of greenhouse gases. We first summarize the nature of the greenhouse effect and then compare its magnitude on these planets.

The temperature of a planet is determined by the requirement that, averaged over time, the infrared emissions to space balance the absorbed solar radiation. The effective radiating temperature of the planet is obtained by equating the thermal emission to that of a blackbody, thus

$$\pi R^2 (1 - A) S_o = 4\pi R^2 \sigma T_e^4 \tag{2.2}$$

or

$$T_e = [S_o(1 - A)/4\sigma]^{1/4} \tag{2.3}$$

where R is the planet's radius, A its albedo, S_o the flux of solar radiation and σ the Stefan-Boltzmann constant.

The difference between the surface temperature and effective temperature of a planet, $T_s - T_e$, is the greenhouse effect due to the gaseous atmosphere and clouds that cause the mean radiating level to space to be at some altitude above the surface.

A quantitative estimate of the greenhouse effect can be obtained under the assumption that only radiation contributes significantly to vertical energy transfer. The Eddington approximate solution of the radiative transfer equation is

$$T_s = T_e(1 + \frac{3}{4}\tau_e)^{1/4} \tag{2.4}$$

with the effective infrared optical thickness τ_e obtained from

$$e^{-\tau_e} = \int e^{-\tau_\nu} B(T_e) d\nu / \int B(T_e) d\nu \tag{2.5}$$

where optical thickness is a dimensionless number such that the fraction $\exp(-\tau_\nu)$ of radiation impinging perpendicularly is transmitted without interaction, B is the Planck blackbody function and the integrations are over all frequencies ν. However, if the atmosphere is sufficiently opaque in the infrared, the purely radiative vertical temperature gradient, dT/dh, may be so large as to be unstable, thus giving rise to atmospheric motions that contribute to vertical transport of heat. In that case a better estimate of the greenhouse warming can be obtained from

$$T_s \sim T_e + \Gamma H \tag{2.6}$$

where H is the altitude of the mean radiation level and Γ is the measured or estimated mean temperature gradient (lapse rate) in the region between the surface and the mean radiating level.[1]

The quantitative theory for the greenhouse mechanism is tested by comparing Mars, Earth, and Venus in Table 2-1. In comparison to Earth, Mars has a small amount of infrared absorbing gases in its atmosphere, while Venus has a dense opaque atmosphere composed mainly of CO_2. The relatively transparent atmosphere of Mars should cause a green-house effect of only a few degrees, as indicated by equation (4). On Earth the lowest few kilometers of the atmosphere are too opaque for pure radiative transfer of heat with a stable lapse rate as a result of the radiative opacity of clouds and the large amount of water vapor in the lower atmosphere. The mean lapse rate in the convectively unstable region is $\Gamma \sim 5.5°C$ km^{-1}, which is less than the dry adiabatic value ($\sim 10°C$ km^{-1}) because of the effects of latent heat release by condensation as moist air rises and cools and because the atmospheric motions that transport heat vertically include not only local convection but also large-scale atmospheric dynamics. The mean radiating level is in the mid-tropo-sphere,[2] at altitude H $\sim$ 6 km. The atmosphere of Venus is opaque to infrared radiation for most of the region between the surface and the cloud tops as a result of CO_2, H_2O, and aerosol absorption. The lapse rate is essentially the dry adiabatic value ($\sim 7°C$ km^{-1}) because of the absence of large latent heat effects and because the large-scale dynamics are unable to produce lapse rates that are appreciably subadiabatic (Stone, 1975). The cloud tops radiating to space are at altitude $H \sim 70$ km.

The observed surface temperatures of Mars, Earth, and Venus are all consistent with the calculated greenhouse warming (Table 2-1). The planets thus confirm the existence and order of magnitude of the green-house effect. These checks of the greenhouse mechanism refer to cases that have had sufficient time to reach thermal equilibrium with space.

Paleoclimate Data

Substantial information exists on the nature of large climate changes that have occurred in the past on Earth. Knowledge is available, for example, of the areas covered by past ice sheets, as evidence by their scouring of rocks; of the areas of sea ice cover, as evidenced by ocean bottom sediments formed by detritus deposited by melting ice; and of oceanic and atmospheric temperatures, as evidenced by the isotopic com-position and geographic distribution of organisms that grew at those times.

Ideally, we would like to have an accurate knowledge of all climate forcings on paleoclimate time scales and of the climate response, which would give us a direct empirical calibration of climate sensitivity. But we do not know all the forcings and the one that is precisely known, variations of solar irradiance due to Earth orbital fluctuations, is a subtle forcing

Table 2-1. Greenhouse Effect on Terrestrial Planets

| Planet | Observed or Estimated | | | | | | T_s(°K) | | |
| | S_o(W m^{-2}) | A | τ | Γ(°C km^{-1}) | H(km) | T_e(°) | Computed | | Observed |
							Eq.(1)	Eq.(2)	
Mars	589	0.15	~0.1	5	1	217	221	222	~220
Earth	1367	0.30	~1	5.5	6	255	293	288	288
Venus	2613	0.75	≳100	7	70	232	≳685	720	~700

S_o = solar irradiance
A = planetary albedo
τ = atmospheric infrared opacity
Γ = atmospheric mean lapse rate
H = mean altitude of emission to space
T_e = effective temperature = $[S_o(1 - A)/4\sigma]^{1/4}$
    (expected planetary temperature in
    the absence of a greenhouse effect)
T_s = surface temperature

Radiative equilibrium
$$T_s = T_e (1 + {}^3/_4 \tau)^{1/4} \qquad (1)$$

Convective equilibrium
$$T_s = T_e + \Gamma H \qquad (2)$$

arising from changes in the seasonal and geographic distribution of radiation rather than a net global change of total incoming flux. Thus, sophisticated models and improved climate data will be needed for the paleoclimate data to yield a direct calibration of climate sensitivity.

However, we can examine the contribution of certain feedback processes to paleoclimate temperature change and thus obtain a valuable measure of the contribution of these feedbacks to climate sensitivity. In particular, the CLIMAP (climate mapping) project (Denton and Hughes, 1981) has determined detailed boundary conditions (see surface temperature, ice sheet coverage and altitude, land boundaries, and sea ice cover) for the Wisconsin ice age (approximately 18,000 years ago).

When the CLIMAP boundary conditions are inserted in a general circulation model (Hansen et al., 1984) they yield a global mean surface air temperature 4°C colder than either today's observed temperature or the temperature produced by the same model with today's boundary conditions. The uncertainty in this cooling is on the order of 1°C and is mainly due to uncertainty in the reconstructed ocean surface temperature, since the model results are fixed closely by the specified boundary conditions.

We can examine sea ice, land ice, and vegetation feedbacks individually by replacing CLIMAP 18K boundary specfications by today's conditions and computing the change in flux at the top of the atmosphere. Since the model sensitivity for a flux imbalance at the top of the atmosphere is known, it is possible to estimate the feedback factor for each of these processes in this way. We present elsewhere (Hansen et al., 1984) a quantitative evaluation of each of these feedback processes based on the CLIMAP boundary conditions.

The sea ice feedback implied by the changes in sea ice coverage during the last ice age is found to be a substantial positive feedback. This conclusion is consistent with the result obtained by the 3-D climate model in the doubled CO_2 experiment described above. This feedback operates on time scales comparable to or longer than the ocean surface temperature response time, and thus it should be included in estimates for the effects of increased CO_2/trace gas abundances on decade-to-century climate change.

The CLIMAP data do not include cloud cover or atmospheric water vapor; therefore, these mechanisms are not tested by the paleoclimate data. However, the total climate sensitivity can be tested empirically on the basis of observed global temperature variations during the past century, as discussed below.

Recent Global Temperature Trends

Surface air temperature observations have been sufficiently widespread to define the global temperature trend for most of the past century

(Hansen et al., 1981). For the same time period, there are substantial observational data on some of the principal global climate-forcing mechanisms. Thus, it is possible to examine the observed trend for the presence of any response to observed forcings, and if it is found, to use this as one calibration of climate sensitivity.

An additional variable is introduced in studying the past century, because for such a short period it can not be assumed that the planet is in thermal equilibrium with space, that is, it is necessary to consider the transient response to variable climate forcing. The response time of surface air temperature to changed climate forcing is dependent on the ocean surface thermal response time, because of the close thermal coupling between the surface air and ocean mixed layer. Although the thermal relaxation time of the mixed layer alone is only several years, exchange of water between the mixed layer and deeper layers may delay the full surface response by decades or even centuries. Thus, the rate of this exchange is the additional variable.

Atmospheric CO_2 has been accurately monitored since 1958, during which time its growth (from 315 ppm in 1958 to 340 ppm today) has been about half of the CO_2 release by fossil fuel burning. If this pattern prevailed earlier, the 1880 CO_2 abundance was about 290 ppm. Possible effects of deforestation make the 1880 abundance uncertain by perhaps 10-20 ppm, but this uncertainty does not greatly influence the analysis of effects on global temperature. The other climate forcing known to be significant during the past century, atmospheric aerosols produced by volcanic emissions, has been reasonably well-defined during this century on the basis of atmospheric transparency measurements. These measurements directly yield the climatically important quantity, visible aerosol optical depth; together with knowledge of intra- and inter-hemispheric mixing times, a useful definition of the volcanic aerosol climate forcing is obtained. Other possible global climate forcings include growth of trace gases, whose effect has become comparable to CO_2 in the past one or two decades but was probably small compared to CO_2 growth earlier (Lacis et al., 1981). Fluctuations in solar output may account for part of the observed climate variability, but adequate measurements are not available for the past century.

The observed global temperature trend for the past century is compared to climate model calculations for two different choices of global climate forcings in Figure 2-6. The observed temperature increase of the past century is matched by the model with CO_2 + volcanoes forcing with an equilibrium climate sensitivity of 2.8°C for doubled CO_2 and an exchange rate $\kappa = 1.2$ cm^2 s^{-1}. This value for κ, the effective vertical diffusion coefficient in the thermocline beneath the 100 m ocean mixed layer, is in the range suggested by current knowledge of oceanic mixing processes.

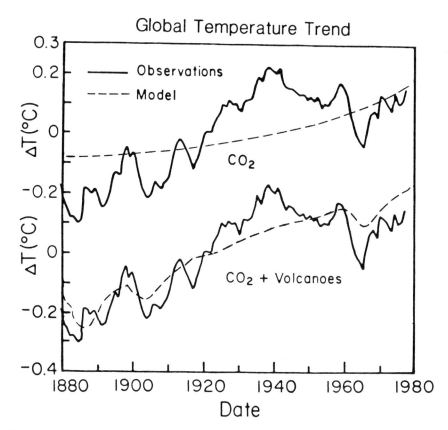

Figure 2-6. Global temperature trend computed with a climate model with sensitivity 2.8°C for doubled CO_2 and an exchange rate $\kappa = 1.2 \, cm^2 s^{-1}$ between a 100 m mixed layer ocean and the thermocline. (After J. Hansen, D. Johnson, A. Lacis, S. Lebedeff, P. Lee, D. Rind, and G. Russell, 1981, "Climate Impact of Increasing Atmospheric Carbon Dioxide," *Science,* **213:**957–966.)

However, if the exchange rate is somewhat larger or smaller, a good fit to the observed temperature trend can still be obtained with a different value for the climate sensitivity. We find that with an exchange rate between the ocean mixed layer and thermocline based on passive tracers ($\kappa \sim 1-2$ $cm^2 s^{-1}$), a climate sensitivity of 2-5°C is needed to provide best fit to the observed global temperature trend.

We conclude that the global temperature trend of the past century is generally consistent with the climate sensitivity estimated by the National Academy of Sciences committees (Charney, 1979; Smagorinsky, 1982): 1.5-4.5°C for doubled CO_2.

CURRENT TRENDS OF GREENHOUSE
GASES AND GLOBAL TEMPERATURE

The average rate of global warming during the past century, 0.04°C per decade, can be expected to increase in the immediate future. This is because the absolute rate of the CO_2 increase (say in ppm/decade) is presently at its highest level and because other trace gases are now increasing at rates that significantly enhance the CO_2 warming.

The CO_2 increase in the decade of the 1970s was about 12 ppm, which was of the order of 25 percent of the total increase of CO_2 for the period 1880-1980. In addition, several other trace gases increased in the 1970s by an amount sufficient to cause a greenhouse warming 50-100 percent as large as that due to CO_2 (Lacis et al., 1981 and Figure 2-7). The rates of change of CH_4 and N_2O were not precisely measured for this period, but Figure 2-7 shows that even with conservative estimates for their growth rates (0.9 percent/yr for CH_4 and 0.2 percent/yr for N_2O), the other trace gases yield a warming 70 percent as great as for CO_2 during the 1970s.

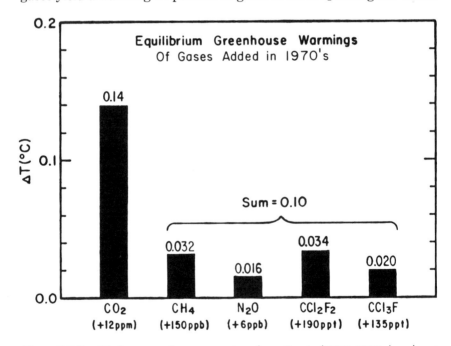

Figure 2-7. Equilibrium greenhouse warmings for estimated 1970-1980 abundance increases of several trace gases, based on climate model with sensitivity ~3°C for doubled CO_2. (After A. Lacis, J. E. Hansen, P. Lee, T. Mitchell, and S. Lebedeff, 1981, "Greenhouse Effect of Trace Gases, 1970-1980," *Geophysical Research Letters* **8:**1035–1038.)

About two-thirds of the chloroflourocarbon increase for 1880-1980 occurred in the 1970s, and it seems likely that the decadel rate of increase of methane was also at a maximum in the last decade.

The eventual warming for these gases added during the 1970s is about 0.2°C if the climate sensitivity is near 2°C for doubled CO_2, but almost 0.4°C if the sensitivity is near 4°C. However, because of the ocean's thermal inertia only some fraction, at most about half, of the warming would be expected to have appeared by the end of the decade.

Natural fluctuations of the smoothed global mean temperature are of the order of 0.1°C for decadel periods. For example, the standard deviation of the 5 year smoothed global temperature in Figure 2-1 is 0.1°C for 10 year intervals. Therefore, although the observed warming in the 1970s (Figure 2-1) is consistent with the increased trace gas abundances, the changes cannot be confidently ascribed to the greenhouse effect.

However, if the abundance of the greenhouse gases continue to increase with at least the rate of the 1970s, their impact on global temperature may soon begin to rise above the noise level. For such a rate of increase the total warming at equilibrium due to gases added in the 1970s and 1980s would be about 0.5°C, for a climate sensitivity of 3°C. Moreover, one would expect that for a 20 year period, a large part of the equilibrium warming would appear by the end of the period. This possible warming should be compared to the standard deviation of observed temperatures, which is about 0.15°C for a 20 year period. This comparison is the basis for anticipating that significant warming is likely to occur by 1990, raising the mean global temperature well above the maximum of the late 1930s (Lacis et al., 1981).

A more quantitative statement about the near-term climate effects of increasing greenhouse gases requires a better understanding of transport and storage of heat in the ocean. This includes the transient response of the ocean to the slowly changing heating pattern at the ocean surface. Realistic treatment will require consideration of the full three-dimensional structure of ocean and atmospheric transports. It will be particularly important to determine the effect of warming and other climate changes at the ocean surface on the ocean mixing and circulation, and thus the ocean feedback on the climate change.

SUMMARY

We conclude that there is strong evidence that a doubling of atmospheric CO_2 will lead to a global warming of at least 1.5°C. Almost all projections of atmospheric composition indicate that an effective doubling of CO_2, including contributions of trace gases, will occur sometime in the next century. Furthermore, for any climate sensitivity in the range $3 \pm 1.5°C$, the global mean warming should exceed natural climate variability during the next one to two decades.

We are left in the very unsatisfactory position of having clear evidence that important climate effects are imminent but not having the knowledge or tools to specify these effects accurately. The principal areas of uncertainty include the equilibrium climate sensitivity, especially the contribution of clouds, and the nature of the transient climate response, which depends on storage and transport of heat by the ocean, including the feedbacks that may occur with changing climate at the ocean surface.

Studies of these components of the climate system are thus suggested as a high priority for research. The chief needs are observational, both global monitoring and local measurements of processes. However, to be effective, such observations must be guided by theoretical studies and modeling. It is particularly important that climate models be developed to reliably simulate regional climate, including the transient response to slowly changing atmospheric composition. This will be difficult because the models need to realistically simulate the effect of greenhouse warming on such factors as standing and transient atmospheric longwaves and ocean currents. The development of such modeling capability will take substantial time and effort, but the benefits from improved understanding of future climate effects will surely warrant the work invested.

NOTES

1. The mean radiating level can be estimated as the averge altitude at which the optical path length of emitted radiation (τ_ν/μ) is unity for the mean cosine of emission angle $\mu \sim 0.5$, with the average over frequencies ν weighted according to the Planck function for the temperature at the emitting level. For Earth the global mean lapse rate is $\Gamma \sim 5.5°C\,km^{-1}$, and based on the spectral computations of thermal emission, the mean altitude of emission to space is $H \sim 6$ km.
2. See note 1.

REFERENCES

Charney, J., chairman, Climate Research Board. 1979. *Carbon Dioxide and Climate: A Scientific Assessment.* Washington, D.C.: NAS Press.

Denton, G. H., and T. J. Hughes. 1981. *The Last Great Ice Sheets.* New York: John Wiley & Sons.

Hansen, J., D. Johnson, A. Lacis, S. Lebedeff, P. Lee, D. Rind, and G. Russell. 1981. "Climate Impact of Increasing Atmosphere Carbon Dioxide." *Science* 213:957-966.

Hansen, J., D. Johnson, A. Lacis, S. Lebedeff, P. Lee, D. Rind, and G. Russell. 1983*a*. "Reply to Technical Comments of MacCracken and Idso." *Science* 220:874-875.

Hansen, J. E., G. Russell, D. Rind, P. Stone, A. Lacis, S. Lebedeff, R. Ruedy, and L. Travis. 1983*b*. "Efficient Three Dimensional Global Models for Climate Studies; Models I and II." *Monthly Weather Review* 111:609-662.

Hansen, J., A. Lacis, D. Rind, G. Russell, P. Stone, I. Fung, R. Ruedy, and J. Lerner. 1984. "Climate Sensitivity: Analysis of Feedback Mechanisms." In *Climate Processes and Climate Sensitivity,* J. E. Hansen and T. Takahashi, eds., Washington, D.C.: American Geophysical Union. pp. 130-163.

Hollin, J. T. 1972. "Interglacial Climates and Antarctic Ice Surges." *Quaternary Research,* 2:401-408.

Lacis, A., J. E. Hansen, P. Lee, T. Mitchell and S. Lebedeff. 1981. "Greenhouse Effect of Trace Gases, 1970-1980." *Geophysical Research Letters* 8(10):1035-1038.

Manabe, S., and R. J. Stouffer. 1980. "Sensitivity of a Global Climate Model to an Increase of CO_2 Concentration in the Atmosphere." *Journal of Geophysical Research* 85:5529-5554.

Manabe, S., and R. Wetherald. 1975. "The Effects of Doubled CO_2 Concentration on the Climate of a General Circulation Model." *Journal of Atmospheric Sciences* 32:3-15.

Manabe, S., and R. Wetherald. 1980. "On the Distribution of Climate Change Resulting from an Increase in CO_2 Content of the Atmosphere." *Journal of Atmospheric Sciences* 37:99-118.

National Academy of Sciences. 1982. *Understanding Climatic Change.* Washington, D.C.: NAS Press.

Smagorinsky, J., chairman, Climate Research Board. 1982. *Carbon Dioxide and Climate: A Second Assessment.* Washington, D.C.: NAS Press.

Stone, P. H. 1975. "The Dynamics of the Atmosphere of Venus." *Journal of Atmospheric Sciences* 32:1005-1016.

Stone, P. H. 1984 (in press). "Feedbacks between Dynamical Heat Fluxes and Temperature Structure in the Atmosphere." In *Climate Processes and Climate Sensitivity,* J. E. Hansen and T. Takahashi, eds., Washington, D. C.: American Geophysical Union, pp. 6-17.

Chapter 3

Estimates of Future Sea Level Rise

John S. Hoffman

INTRODUCTION

Accurate monitoring of atmospheric carbon dioxide began 25 years ago. Since then, sufficient scientific evidence has been developed for two National Academy of Sciences review panels to conclude that sometime in the next century, atmospheric concentrations of CO_2 will almost certainly double and raise the atmosphere's mean surface temperature by at least 1.5°C (2.7°F) and possibly as much as 4.5°C (8.1°F) (Charney, 1979).[1] Such a warming should also raise the global sea level. While adequate knowledge of the various determinants of sea level rise has been developed, until now this diverse knowledge has not been used to estimate possible sea level rise trends.

This chapter presents a range of sea level rise estimates, termed scenarios, that were developed on the basis of knowledge collected from a variety of disciplines, including energy economics, geochemistry, biology, atmospheric physics, oceanography, and glaciology. The most restrictive assumptions from these disciplines were linked together to generate a "conservative" scenario, which projects a sea level rise of 56.2 cm (22 in) by 2100. The least restrictive assumptions were used to generate a "high" scenario, which projects a rise of 345 cm (11.5 ft) by 2100.[2] Two mid-range scenarios were also developed, the mid-range low scenario which projects a rise of 144 cm (4.8 ft) and the mid-range high scenario which projects a rise of 216 cm. (7 ft). In the author's judgment, future sea level rise is most likely to fall in this range.

Although the scenarios span a wide range of sea level rise, they can still be used in analyzing environmental and economic impacts and to evaluate options for preventing or mitigating the adverse effects of this phenomenon. Narrowing the range of estimates of sea level rise would, of course, make these tasks easier. But rapid progress will be made only if funding is increased for key scientific disciplines. Thus, policy makers in government and business may also wish to use the scenarios in making their own assessments of the economic value of developing more precise sea level rise estimates.

THE APPROACH USED FOR ESTIMATING SEA LEVEL RISE

Future global sea level will depend primarily on three factors: the total quantity of water filling the oceans' basins; the temperature of the oceans' layers, which determines the density and volume of their waters; and the bathymetry (shape) of the ocean floor, which determines the water-holding capacity of the basins. A rise in global temperature can, by a variety of physical mechanisms, transfer snow and ice from land to the sea, increasing the quantity of water in the ocean basins, and can raise the oceans' temperatures, causing the thermal expansion of their volumes. Changes in the bathymetry of the oceans' floors occur independently of climate change. Because geological changes in the ocean floor could not raise or lower global sea level by more than a centimeter or two by 2100 (Clark et al., 1978), this factor was not considered in constructing the global scenarios. An evaluation of the impacts of sea level rise at specific coastal sites, however, will require consideration of local uplift or subsidence, which by 2100 could cause changes in land elevation that are large enough to be of significance to local planning (Boesch, 1982).

Projecting sea level rise requires the means to estimate future changes in atmospheric composition, to relate these changes to global warming, and then to determine how the warming can cause land-based snow and ice to enter the sea and the oceans to expand thermally. Figure 3-1 summarizes these relationships and the various alternative assumptions and models used to represent them. Details about these relationships, models, and assumptions are discussed below.

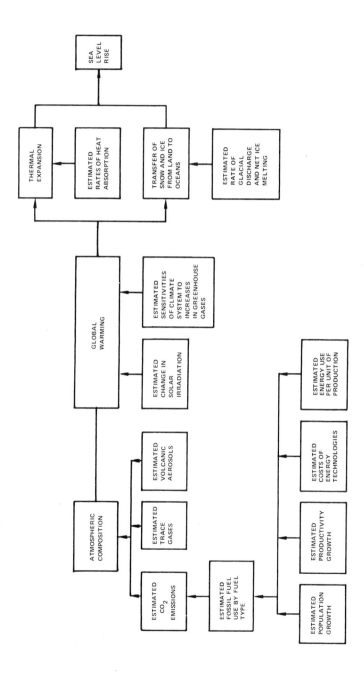

Figure 3-1. Conceptual basis for estimating sea level rise. High, low, and mid-range assumptions were made or derived from models about each factor that determined atmospheric composition, global warming, thermal expansion, ice and snow discharge, and thus sea level rise. By spanning a full range of estimates for these factors and coupling them on a yearly basis, conservative, mid-range, and high scenarios were generated on a yearly basis to the year 2100.

FUTURE ATMOSPHERIC COMPOSITION

This section provides further background on the models and assumptions used to estimate CO_2 levels, one of the most important determinants of global temperature. The models and assumptions used to estimate the growth of atmospheric concentrations of trace gases that will influence climate are also discussed.

Estimates of Increases in Atmospheric Concentrations of CO_2

Although large quantities of carbon circulate between the oceans, atmosphere, and biosphere, before the widespread use of fossil fuels and the removal of forests, the cycling of carbon among these natural compartments had evolved to maintain a more or less stable level of atmospheric CO_2. Ice cores, for example, provide evidence that CO_2 levels fluctuated no more than 40 ppm for thousands of years (Barnola et al., 1983).

The industrial revolution brought about an enormous change in the carbon cycle, altering its balance dramatically. Machines helped expand the economy to unprecedented levels, allowing much greater amounts of "work" to be done than in pre-industrial economies. In accomplishing this feat, however, enormous quantities of gas and oil had to be used. Combustion of these fuels has grown so large that by 1980, 5 billion metric tons of carbon were being released into the atmosphere every year (Rotty, 1983).

Because this fossil carbon had rested inertly in the ground for hundreds of millions of years (outside the earth's normal cycling of carbon), the sudden injection of CO_2 into the atmosphere has greatly overwhelmed the capacity of the biosphere and oceans to absorb CO_2. In essence, each year's fuel combustion restores a quantity of carbon to the atmosphere that had taken plants thousands of years to remove. Consequently, atmospheric concentrations of CO_2 have been increasing rapidly: since the 1860s, concentrations are estimated to have risen 20 percent worldwide, while since 1957, monitored data show an increase in atmospheric CO_2 concentrations of more than 7 percent (Keeling et al., 1976). Future levels of CO_2 will depend on future emissions and on the future capacity of the earth's oceans and biosphere to absorb emissions rather than having them remain in the air—the so-called fraction airborne of CO_2.

Future Carbon Dioxide Emissions. Future levels of CO_2 emissions will be determined by economic growth and technological change. Economic growth is driven by population and productivity growth. The development of technologies for energy production and use is also the consequence of a complex set of factors. In this analysis, CO_2 emissions were projected by using a world energy model developed at the Institute for Energy Analysis (IEA) and a series of assumptions gathered from a review of the literature. This model determines fuel use and emissions by simulating market mechanisms. Because many factors about the relevant variables are uncertain, various assumptions about these factors can be run in the model to produce internally consistent estimates of fuel use (Edmonds and Reilly, 1981).

The IEA model disaggregates the world into nine regions, balancing the demand and supply of energy in each region by internal production and external trading, in which production and transportation costs determine the costs of satisfying a region's demands. Once fuel use trends are generated, standard emission coefficients are used to estimate CO_2 emissions.

Model validity was established in a number of ways, including a parametric sensitivity analysis of its various coefficients. The model's outputs were also compared with work done elsewhere, such as that reported in IIASA's (the International Institute for Applied Systems Analysis) *Energy in a Finite World* (Haefle, 1981), and provide consistent, although somewhat lower estimates of energy use than many other efforts.[3] The IEA model, however, has the advantage of being able to test the implications of a wide range of assumptions about energy technologies, policies, and events, as well as the sensitivity of fuel use to other assumptions.

Using this model, a detailed study has been conducted of an array of policies intended to reduce CO_2 emissions (Seidel and Keyes, 1983). The study indicates that CO_2 emissions are likely to grow regardless of the policies implemented. For example, a tax that quadrupled fossil fuel prices by 2000 reduced CO_2 emission growth insufficiently: a 2°C global warming was delayed by only five years. Thus, the world's capability to displace fossil fuels in an effective manner appears quite limited.

Some of the assumptions used in this effort were quite conservative, even for the high scenario. For example, for population growth, a single demographic projection was used: all regions of the world were assumed to reach zero population growth by 2075. Many demographers would regard such an assumption as optimistic. Nonetheless, based upon the work of Keyfitz et al. (1983) and others, it seemed reasonable.

For productivity growth, two assumptions were made:

For the high scenario, world labor productivity growth was assumed to start at 3.5 percent per year and decline linearly to 2.2 percent by 2100.
For the low scenario, world labor productivity growth was assumed to start at 2.2 percent and decline linearly to 1.7 percent by 2100.

Both the high and low assumptions are below the rate of productivity growth achieved in the world in the last 30 years.

Next, the energy used for production had to be estimated. This use will depend on the mix of goods and services produced in the world's economies and the technologies utilized to meet those needs. For all scenarios, the energy use per unit of output was assumed to decline with economic growth, reaching 40 percent of its current level by 2100. This assumption was based on the expectation of increasing conservation and the tendency of societies growing more affluent to shift output toward less energy-intensive products (Edmonds and Reilly, 1981).

The selection of future energy sources to meet total energy demand was based on the comparative costs of satisfying the energy needed, which depends on the availability of resources, the costs of transporting fuels, and on available energy extraction and use technologies. Three sets of assumptions were made regarding these areas.

For the high scenario, a best guess based on current research was made about the future costs of various energy technologies. Progress was assumed for emerging technologies such as solar, but no unprecedented breakthroughs were assumed.

For the mid-range scenarios, emissions were assumed to be halfway between the conservative and high scenarios.

For the conservative scenario, the cost of nuclear energy was halved, starting in 1980, from current levels, thereby increasing the comparative attractiveness and use of this form of energy. Such an assumption obviously is unrealistic, but by lowering the cost of a nonfossil fuel energy so quickly, it created a more restricted fossil fuel growth.

The IEA model, of course, was used to integrate these different assumptions and produce various emissions trends.

Fraction Airborne Assumptions. A precise accounting of all the sources and sinks of carbon has not yet been achieved. This alone makes precise specification of the fraction of CO_2 that will remain airborne impossible. In addition, many of the biogeochemical processes that control the exchange of carbon may be somewhat influenced by climate and rising CO_2, thus making estimation of the future fraction airborne even more difficult.

Nevertheless, knowledge of the carbon cycle is sufficient to make some reasonable limiting assumptions. Oceans (the primary sink for emissions), for example, will have limited capacity to absorb CO_2, and the implication is that the percentage of future emissions absorbed by that sink will decline (Broecker et al., 1980). Three assumptions were used to represent the future evolution of the fraction airborne.

For the conservative scenario, the best estimate of the historic fraction airborne was used (53 percent) as an estimate of its future value.

For the high scenario, a carbon cycle model that considers some, but not all, of the factors that will cause the fraction airborne to rise was used to generate an evolving and rising fraction airborne (from 61 percent to 80 percent by 2100).

For the mid-range scenarios, the fraction airborne was assumed to rise as in the carbon cycle model.

The choice of 53 percent as the the most restrictive assumption, rather than a lower estimate advocated by a few researchers (e.g., 40 percent), can be justified for two reasons. First, the fraction airborne will almost certainly rise, making it almost certain that over the time period of concern that the fraction airborne would actually rise above the low estimate. Second, evidence has accumulated that makes earlier claims of large amounts of deforestation (Woodwell, 1978) seem too high. A higher estimate of deforestation implies greater CO_2 emissions and thus more CO_2 remaining airborne to produce the observed increase in atmospheric concentrations. Biologists (Lugo, 1980) have shown that early estimates of forest contribution ignored forest regrowth (which would recycle the carbon back into biomass), while oceangraphers (Broecker et al., 1980) have shown that the ocean could not physically absorb the amounts of CO_2 required by the deforestation estimates. Thus, while some doubt still persists about the best estimate of the historical fraction airborne, the use of a constant 53 percent for the future fraction airborne probably will, in any case, be a conservative assumption of its future level. If, however, Woodwell turns out to be correct, from 1990 to 2050, the additional CO_2 resulting from deforestation will compensate for the lower faction airborne, and the scenarios discussed here will be valid. After 2050, however, forests would be exhausted and global heating and sea level rise would not increase as quickly as predicted (Keyes et al., 1984).

While the model of the carbon cycle developed by Emmanuel and Killough of Oak Ridge National Laboratories (ORNL) predicts a fraction airborne that rises in the high scenario, it does not represent the highest plausible level of CO_2 remaining in the air. The model simulates the exchange of carbon from oceans, biota, and the terrestrial features of the earth and the atmosphere at a highly aggregate level and in a simplified manner (Killough and Emmanuel, 1981). As such, the model represents some of the physical processes of the carbon cycle such as the oceans' limited capacity to absorb CO_2, that could raise the fraction airborne. The model does not, however, consider possible changes in oceanic circulation that could be induced by warming. As the surface layers of the ocean warm, circulation could decrease, further reducing the oceans' capacity to absorb CO_2. Thus, although the model's outputs were used as the high assumption, the ORNL model appears to be a more realistic assumption for the fraction airborne than the historical estimate and was used in the mid-range estimates.

Increases in Atmospheric Concentrations of Trace Gases

During the 1970s the trace gases methane, nitrous oxide, and chlorofluorocarbons began to be monitored more accurately. Observed increases in the concentrations of these gases appear to have had between 50 and 100 percent of the warming effect of the rise in CO_2 in that decade (Lacis et al., 1981). Unfortunately, atmospheric concentrations of the trace gases could not be projected in the same manner as concentrations of CO_2. With the exception of the chlorofluorocarbons, our knowledge of man-made and natural sources and sinks of these gases, the biogeochemical exchange mechanisms between their storage compartments, and the atmospheric chemistry that governs their chemical form in the atmosphere has not yet advanced to the point that their future concentrations can be meaningfully projected by estimating emissions and fraction airborne separately. Thus, for methane and nitrous oxides, direct historical observations of the concentrations of these trace gases in the atmosphere were used as a basis for estimating future concentrations; for chlorofluorocarbons, emissions were projected and a simple model of their expected residence time in the atmosphere was used.[4]

For the low scenario, atmospheric concentrations of chlorofluorocarbons were assumed to increase based on 60 year and 120 year half-lives for $CFCl_3$ (R-11) and CF_2Cl_2 (R-12), respectively, and on the assumption that emissions increase at 0.7 percent of the 1980 level every year until they are capped in 2020 (Gibbs, 1983); nitrous oxide concentrations were assumed to grow at 0.2 percent per year (Weiss, 1981); and methane concentrations were assumed to grow at 1 percent per year (Rasmussen, 1981).

For the mid-range scenarios, chlorofluorocarbon emissions were assumed to grow annually by 2.5 percent of the 1980 level until they are capped in 2020 (Gibbs, 1983); the concentrations of atmospheric nitrous oxide were assumed to grow by 0.45 percent per year (Weiss, 1981); and the concentrations of methane were assumed to grow by 1.5 percent per year (Rasmussen, 1981).

For the high scenario, chlorofluorocarbon emissions were assumed to grow annually by 3.8 percent of the 1980 level until they are capped in 2020 (Gibbs, 1983); nitrous oxide concentrations were assumed to grow 0.7 percent per year (Weiss, 1981); and methane concentrations were assumed to grow 2.0 percent per year (Rasmussen, 1981).

The possibility exists that new sinks or new sources of methane and nitrous oxide could arise, either as a result of climatic change or human activity. For example, global warming could result in methane hydrates located on northern continental shelves becoming a source of methane emissions (McDonald, 1982). Rising CO_2 could cause changes in plant and soil interactions (through the fertilization effects of CO_2), making soils a greater source of nitrous oxide (Lemon, 1983). Similarly, the possibility exists that current sources or sinks may become exhausted. If either of these situations arises, the resulting change in the biogeochemical cycle could radically alter the growth of atmospheric concentrations of these gases. Unfortunately, without a significant increase in research in these areas, it will be difficult to improve projections of the future concentrations of these gases.

FUTURE GLOBAL TEMPERATURE

Global surface temperature is determined by the radiation received from the sun (mainly in the visible part of the spectrum), the reflectivity of the earth's surface and atmosphere, and the amount of outgoing invisible infrared radiation that the atmosphere traps. Without the atmosphere's greenhouse constituents, the thermal balance (temperature) of the earth would be relatively straightforward to estimate: the outgoing infrared radiation would balance incoming visible radiation at an effective temperature of approximately $-18°C$ ($0°F$) (Hansen et al., 1983).

Fortunately for the existence of life on earth, the atmosphere has greenhouse gases, such as H_2O, CO_2, methane, chlorofluorocarbons, N_2O, and other trace gases, which trap some of the escaping infrared radiation, thereby raising the earth's temperature. These gases are called greenhouse gases because they allow sunlight to pass unimpeded to the earth's surface but absorb the infrared radiation given off by the earth. After absorbing the infrared energy, the gases warm, and then reradiate the energy, half of which goes downward to the planet's surface. With the earth's current concentrations of greenhouse gases, enough energy is trapped to raise the earth's surface temperature to $33°C$ ($58°F$) (Hansen et al., 1983).

Increasing concentrations of atmospheric CO_2 and other infrared-absorbing gases will cause more radiation to be trapped and raise temperatures. The temperature rise associated with the radiative effect of rising concentrations of greenhouse gases would not be difficult to estimate if the warming did not further alter the composition of the earth's atmosphere or change the earth's reflectivity: a doubling of CO_2 would raise the average surface temperature 1.2°C. In reality, however, the radiative forcing (that is, the initial warming) will change the atmospheric composition and surface reflectivity of the earth in ways that amplify the warming. Determining the magnitude of amplification is difficult, thus adding uncertainty as to the effect of rising CO_2 (or other greenhouse gases).

The Amplification of Initial Warming Effects

The initial warming will cause water vapor to evaporate; as the evaporated water enters the air, it will trap more infrared radiation, thereby causing further temperature increases. The initial warming will also melt snow and sea ice, thereby reducing the sunlight reflected back into space and causing even further warming. Other feedbacks also exist; changes may occur in cloudiness or cloud height, for example, which, depending on how they occur, could raise or lower the amplification. Unfortunately, at this time, scientific understanding of all the relevant climatic processes is insufficient to determine the total amplification with precision. The best the scientific community has been able to do is to conclude that the total amplification will be at least 25 percent (Charney, 1979).

In building the scenarios, uncertainties about the response of the various parts of the climate system were encompassed by using a wide range of the estimates of thermal sensitivities (or thermal equilibrium responses) made by the National Academy of Sciences (NAS) after two thorough analyses of the evidence.[5]

For the low scenario, a 1.5°C temperature rise was assumed for a doubling of CO_2 concentrations.
For the mid-range scenarios, a 3.0°C rise was assumed for a CO_2 doubling.
For the high scenario, a 4.5°C rise was assumed for a doubling of CO_2.

Other Forcings

Two other forcings can influence the earth's thermal balance in the time frame under consideration: changes in particulates and volcanic aerosol levels and changes in solar irradiation. Shifts in solar irradiation would be caused by shifts in the output of energy from the sun or by alterations of the earth's orbit or axis. These latter changes occur far too slowly and are not relevant within the time frame under consideration (Hays et al., 1976). Thus, the possibility of an ice age driven by the Milankovitch effect may be ruled out; the time frame for such a cooling would be much longer than the 120 years considered in this effort.

Preliminary evidence from NASA and other astronomical observations, however, indicates that yearly changes in solar irradiation could be large if the changes recorded to date continue in a single direction for long periods of time (Wilson, 1982). However, no evidence for any systemic unidirectional change exists at present (Wilson, 1982).

Shifts in concentrations of tropospheric particulates can be produced by economic activity. The warming or cooling effects of these particulates, however, vary with their physical nature and their geographic distribution (Bach, 1981). At this time, it appears that both their positive and negative thermal effects should cancel out (Bach, 1981). Therefore, tropospheric particulates were assumed to have no net effect. Volcanic eruptions, however, can transport aerosols to the stratosphere, which will lead to a cooling effect. Agung, a large Indonesian volcano, exploded in 1963, for example, decreasing global temperatures for the following year (Hansen et al., 1978).

Because of their potential influences, changes in volcanic activity and solar irradiation were considered in several "special case" scenarios. In one scenario it was assumed that throughout the next 100 years, the optical opacity of the atmosphere (the factor that volcanic aerosols increase) would remain at the level it reached during the two decades of the last century when optical opacity was highest. Changes in solar irradiation were also tested in several special case scenarios. In some of these scenarios, a linear increase to a total change of 0.5 percent was assumed over the next 100 years, while in other scenarios, solar irradiation was assumed to decrease linearly by that same amount. Evidence supporting either of these two shifts, however, is lacking and these scenarios should not be looked upon as accurate predictors.

Atmospheric Composition and Temperature Sensitivity

A complex climate model was not used to incorporate the temperature effects of the forcings on yearly global temperature. Instead, an equation was used (see Figure 3-2) that had been empirically fit to a one-dimensional radiative convective climate model (Hansen et al., 1981). This equation allowed the various changes in the independent factors, or forcings (e.g., changes in atmospheric composition or solar irradiation), to be related to changes in atmospheric composition and to assumptions about thermal equilibrium, the levels of the aerosols, and solar irradiance. The equation had been coupled to a box-diffusion model of the ocean so that it actually generates a heat flux to the ocean, the size of the flux depending on the difference in temperature between the air and the ocean's surface. In this way, the various forcings and responses of the climate system were integrated with an "oceans model" that simulates the delays

$$F(t) = \frac{(2.6 \times 10^{-5})\Delta C}{[1 + (2.2 \times 10^{-3})\Delta C]^{0.6}} - \frac{5.88}{T_e} + 10^{-3}(\Delta T) + \frac{3.685}{T_e^2} \times 10^{-4}(\Delta T)^2$$

$$- \frac{(4.172 \times 10^{-7})}{T_e}(\Delta C)(\Delta T)$$

$$+ 1.197 \times 10^{-3}(\Delta F_4)^{0.5} + 5.88 \times 10^{-3}(\Delta N)^{0.6} + 3.15 \times 10^{-4}(\Delta F_3)$$

$$+ 3.78 \times 10^{-4}(\Delta F_2) - 1.197 \times 10^{-4}(\Delta F_4)(\Delta N) - 2.4 \times 10^{-2}(\Delta V)$$

$$- 2.1 \times 10^{-3}(\Delta V)^2 - \frac{1.17 \times 10^{-3}(\Delta T)(V)}{T_e} + 3.184 \times 10^{-1}(\Delta S)$$

where:

ΔC = change in CO_2 from 1880

ΔT = change in temperature from 1880

T_e = thermal equilibrium

F_4 = methane variable

N = N_2O variable

F_3 = CCl_3F variable

F_2 = CCl_2F_2 variable

V = volcanic variable (optical depth)

S = solar variable, the change in solar luminosity from its equilibrium value divided by its equilibrium value

Figure 3-2. Flux equation used to couple atmospheric forcings, choice of thermal equilibrium, and box diffusion model of heat transport into the oceans. *(From J. Hoffman, D. Keyes, and J. Titus, 1983, Projecting Future Sea Level Rise: Methodology, Estimates to the Year 2100, and Research Needs, 2nd rev. ed., U.S. GPO No. 055-000-00236-3. Washington, D.C.: Government Printing Office.)*

that will occur in atmospheric warming as a result of the ocean's capacity to absorb heat. The heat flux is generated on a year-by-year basis, as atmospheric concentrations of greenhouse gases change. Thus, atmospheric and oceanic temperatures rise slowly, with heat gradually being passed to lower layers of the ocean.

FUTURE OCEAN AND GLACIAL
RESPONSES TO GLOBAL WARMING

The effects of global warming on sea level also depend on how that warming influences the oceans and ice sheets of the world. This section describes the models and assumptions used to produce the scenarios.

Ocean Response

As discussed, increases in global temperature will not raise the average temperature of all the oceans' various layers immediately. The temperature of the surface waters will respond most quickly, essentially increasing in synchrony with atmospheric temperature. The transport of heat downward, however, will be slower (Charney, 1979). Furthermore, as an atmospheric warming occurs, the circulation of ocean waters will probably decrease, slowing the formation of deep cold water at the poles and thus the upwelling of cold water elsewhere. Changes in circulation could ultimately alter the rate of oceanic heat absorption, especially toward the end of the next century. Because the speed at which these circulation changes will occur is not clear, however, this possibility was not considered in creating the scenarios. The effects of such a change on thermal expansion would in any case be ambiguous. A decline in heat absorption would tend to slow thermal expansion, but this tendency would be counteracted by the fact that global warming would occur faster.

The ocean model assumes that a column of water can be used to represent the oceans and that heat can be transported downward like a passive tracer. (Figure 3-3 shows the equations used.) The estimate of the rate of heat diffusion downward uses data from various studies of radioactive and chemical tracers. Heat was not assumed to be transported below the mixed layer of the ocean (Hansen et al., 1981). In the model used, the mixed layer has a depth of 100 m (328 ft) and the thermocline has a depth of 900 m (2,952 ft). The mixed layer temperature was assumed to be independent of depth. A diffusion equation with a constant thermal conductivity determines the thermocline's temperature.

For the low scenario, the diffusion coefficient assumed was $1.18 \text{ cm}^2 \text{ sec}^{-1}$.

For the mid-range scenarios, the diffusion coefficient was assumed to be $1.54 \text{ cm}^2 \text{ sec}^{-1}$.

For the high scenario, the diffusion coefficient assumed was $1.9 \text{ cm}^2 \text{ sec}^{-1}$.

The heat flux is estimated for semi-monthly time periods. The appropriate ΔT for calculating $F(t)$ in each time period $(t = n)$ is the value estimated for the previous time period. For a simple one-layer ocean model, ΔT is obtained by solving the following differential equation:

$$\frac{d\Delta T}{dt} = \frac{F(t)}{Co}$$

where Co is the heat capacity of the ocean per unit area (cal cm^{-2}). The temperature change in the mixed layer (ΔTm) is a solution of the equation:

$$cHm \frac{d\Delta Tm}{dt} = F(t) - F_D(t)$$

where:

c = heat capacity of the water
Hm = depth of mixed layer
$F(t)$ = heat flux from the atmosphere
$F_D(t) = \lambda \left. \frac{\partial \Delta T}{\partial Z} \right| Z = Hm$ is the heat flux into the mixed layer

Note that the Z axis is directed toward the bottom of the ocean. Since g, cm, sec, cal were used, the heat conductivity ∂ is numerically equal to the heat diffusivity.

Temperature change in the thermocline is determined by the diffusion equation:

$$\frac{c\partial \Delta T(Z,t)}{\partial T} = \kappa \frac{\partial^2 \Delta T(Z,t)}{\partial Z^2}$$

the boundary conditions for ΔT are:

$$\Delta T = \Delta Tm \text{ at } Z = Hm$$

and zero heat flux at the bottom of the thermocline:

$$\frac{\lambda \partial \Delta T}{\partial Z} = 0 \text{ at } Z = H + Hm$$

Note that ΔTm and ΔT are temperature changes between the time 1880 and t.

Figure 3-3. Equations used in ocean model for heat transport. *(After J. E. Hansen, D. Johnson, A. Lacis, S. Lebedeff, D. Rind, and G. Russell, 1981, "Climatic Impacts of Increasing Atmospheric Carbon Dioxide," Science* **213**:957-966.)

The thermocline has a depth of 900 m (2,952 ft). The mixed layer temperature was assumed to be independent of depth. A diffusion equation with a constant thermal conductivity determines the thermocline's temperature.

For the low scenario, the diffusion coefficient assumed was 1.18 cm^2 sec^{-1}.

For the mid-range scenarios, the diffusion coefficient was assumed to be 1.54 cm^2 sec^{-1}.

For the high scenario, the diffusion coefficient assumed was 1.9 cm^2 sec^{-1}.

This range of diffusion coefficients assured that neither too little nor too much heat was absorbed by the oceans. Thermal expansion was calculated by using mean temperatures, salinities, and pressures for each layer in the water column, so that for each layer a standard coefficient of expansion could be used once the change in temperature was ascertained. This estimating approach slightly mis-estimates actual thermal expansion because actual ocean temperatures and fluxes vary latitudinally. An analysis of this averaging error, however, indicates that it is not large.[6] The approach generates a good first order estimate of thermal expansion.

Snow and Ice Contribution to Sea Level

The amount of water locked on land areas in the form of snow and ice will change as global temperatures increase. Large amounts of such snow and ice exist on Greenland and Antarctica. Because polar temperatures are expected to rise more than the global average (Manabe, 1983), the direct effects of warming on increased melting, evaporation, and sublimation will be great; a less direct effect will be the influence of warming on deglaciation. The net effects of these decreases in land-based snow and ice will be reduced somewhat by increases in snowfall that can be expected in polar areas. Warmer air will tend to carry more moisture and snow to extreme latitudes.

It is difficult at this time to determine with certainty the net effects of increased melting and increased snowfall on mass balance. An initial examination of doubled CO_2 experiments using the general circulation model developed at the Goddard Institute for Space Studies (GISS) reveals that in situ melting (for an equilibrium temperature at CO_2 doubling) could be the equivalent of 10.5 to 16.5 mm (0.4 to 0.6 in) per year of sea level rise. These estimates are subject to some overestimation for a number of reasons (Hoffman et al., 1983). For example, not all meltwater will run off; some will percolate into the ice sheets. This meltwater will cause additional crevassing and ice softening, which, in turn, could accelerate the deglaciation of the ice sheets.

Deglaciation is likely because large parts of the East and West Antarctic icefields are grounded below sea level, and thus are subject to rapid collapse. With the reduction of sea ice and the warming of the polar oceans, these areas are much more likely to experience future deglaciation. Some of these icefields, notably in the West Antarctic, are held in place by pinning grounded below sea level. At present, these pinnings prevent the glaciers from moving rapidly seaward. Warmer oceanic waters may melt and remove the pinnings, soften the ice, and thus lead to a much more rapid movement of the ice. Ultimately, large portions of the land-based ice sheets could enter the ocean. The speed of deglaciation, if it occurs, will depend upon many things: ocean currents, the frictional coefficients of the "surging" ice, and the specific topography of pinnings and outlet channels (Bentley, 1983; Hughes, 1983).

Evidence exists that the West Antarctic completely disappeared during previous global warmings, raising sea level by 5-6 m (16-20 ft) 120,000 years ago (Mercer, 1978). Unfortunately, little scientific effort has been expended to determine the rate of deglaciation in the near future. Two estimates have been published that speculate that the earliest time of the total collapse of the West Antarctic icefields would be 200 and 500 years from now (Hughes, 1983; Bentley, 1983). Unfortunately, neither researcher made interim estimates of deglaciation. For icefields in East Antarctica or Greenland, no estimates of deglaciation have been made at all, despite their potential vulnerability.

Given the absence of studies, it was impossible to use process models of icefields to estimate the deglaciation that global warming could cause. EPA is now supporting very limited research in this area for the second round of sea level scenarios, due to be completed in late 1984. That effort will use process models to examine key physical processes, boundary conditions, parameters and their potential evolution, in order to create better scenarios of glacial contributions under various scenarios of global temperature rise. In the present effort, however, methods that are far less reliable had to be used.

One possibility was to assume that the meltwater estimated by the GISS model for a CO_2 doubling will produce an equivalent sea level rise through either of two mechanisms: either by directly entering the sea as runoff or by indirectly causing faster deglaciation as the refreezing process changes ice flow characteristics. Because the GISS model output provides estimates of melting only for the equilibrium warming (4.1°C) for a doubling of CO_2, estimates of the total melting that would take place over the next century, when temperatures would at first be lower and then warmer than the doubled temperature, could not be made directly from the model output. A relationship between global temperatures and the melting thus had to be assumed. Assuming proportionality of melting to global warming, a sea level rise of roughly 75-112 cm (2.5-3.7 ft) can be forecast by 2100 (assuming no error in the GISS estimates).

Another method for predicting the contribution of snow and ice transfer from land to sea is to assume a continuation of the past association between thermal expansion and total sea level rise. Because part of the historical sea level rise can be attributed to thermal expansion, estimates of the ratio of thermal expansion to snow/ice are possible. By extrapolating that ratio into the future, scenarios of snow/ice contributions can be generated. Since different estimates of past sea level rise exist, this approach requires generating both a high and conservative snow/ice thermal expansion ratio.

The sea level rise estimates used in this process were based on work by Barnett (1983) and Gornitz et al. (1982), who estimated historical sea level rises of 10-15 cm (4-6 in). The Gornitz group also estimated that thermal expansion accounted for 5 cm (2 in) of the rise; thus, depending on whose global sea level rise estimate is used, there is either a 5 or 10 cm (2 or 4 in) residual. This latter approach was the one actually used to generate the scenarios. Using these values, two assumptions were made:

For the low scenario, the ratio of future snow/ice contribution to future thermal expansion was assumed to be "one to one."
For the high scenario, a "two to one" ratio was assumed.
For the mid-range-low scenario, the low snow and ice assumption was used; for the mid-range-high scenario, the high snow and ice assumption was used.

The ratio approach has serious flaws. At best, it has a weak physical basis. It relies on estimates of past temperature change that are somewhat uncertain. Furthermore, it also extrapolates a constant ratio of snow and ice to thermal expansion. If alpine glaciers were the source of the sea level rise not explained by thermal expansion, the possibility exists that these sources may become exhausted towards the end of the forecast period, reducing the ratio over time. If deglaciation were the source, its many nonlinear features could lead to underestimates by the end of the forecast period. Clearly, more research is needed on this problem; the responses of land-based ice should be made on the bases of direct projections of snow and ice contributions with process models of deglaciation melting, and runoff from all snow and ice fields. Nevertheless, the flaws in the estimating approach are not egregious enough to invalidate the effort. The approaches used to estimate sea level rise constitute an attempt to address this source of sea level rise in a reasoned manner and appear to be a far better choice than ignoring the possible contribution of snow and ice resting on land, particularly given the clear importance these sources should have for future sea level rise. The estimating procedure, while less than perfect, at least starts to utilize the existing base of knowledge to estimate sea level rise.

FUTURE SEA LEVEL RISE

Considering only changes in greenhouse gases (not the special case scenarios that deal with other forcings), sea level could rise as much as 345 cm (136 in) and as little as 56.2 cm (22 in) (Hoffman et al., 1983) by 2100. Neither of these extreme estimates is likely, however, since the probability of all the conservative or all the high assumptions turning out to be true is very small. The moderate thermal expansion scenario, because it assumes either the middle ground for all assumptions or the assumption that appeared most realistic, constitutes a much more likely trend for this component of sea level rise. Two scenarios were produced for snow and ice contribution. For the moderate scenario, the low snow and ice ratio was assumed. For the mid-range-high scenario (not discussed in other chapters), the high snow and ice ratio was used. The moderate scenario produces a rise of 144.4 cm (4.8 ft) by 2100, while the mid-range-high scenario produces a rise of 216 cm (7 ft) by 2100.

Table 3-1 summarizes the changes by quarter century, for the conservative, moderate, mid-range-high and high scenarios. Extrapolations of the historical rate of rise are included for comparison. Table 3-2 summarizes some of the special case scenarios that considered changes in volcanic activity and solar irradiation.

REDUCING UNCERTAINTIES

In all foreseeable circumstances, sea level is likely to rise by amounts considerably greater than this past century's rise. The most conservative assumptions used in this analysis lead to an accelerating sea level rise and a total rise that is 400 percent greater than that of the last 100 years. Nevertheless, many uncertainties exist about the rate of rise. The very high scenario has over seven times the sea level rise of the conservative scenario. Part of the variance between scenarios may be an artifact of the relatively crude methods used for estimating sea level rise, rather than a lack of insight into its physical mechanisms. That part of the uncertainty can probably be eliminated in the next two years if more resources are to devoted to the estimating effort. However, even those improvements may not provide more precise estimates.

In order to improve substantially the estimates of future sea level rise and to narrow the range of scenarios, more time and more scientific

Table 3-1. Estimated Sea Level Rise, 2000–2100, by Scenario (in cm, with inches in parentheses)

Year	Conservative	Mid-Range Scenarios Moderate[a]	High	High Scenario	Historical Extrapolation
2000	4.8	8.8	13.2	17.1	2–3
	(1.9)	(3.5)	(5.2)	(6.7)	(0.8–1.2)
2025	13.0	26.2	39.3	54.9	4.5–8.25
	(5.1)	(10.3)	(15.5)	(21.6)	(1.8–3.2)
2050	23.8	52.3	78.6	116.7	7–12
	(9.4)	(20.6)	(30.9)	(45.9)	(2.8–4.7)
2075	38.0	91.2	136.8	212.7	9.5–15.5
	(15.0)	(35.9)	(53.9)	(83.7)	(3.7–6.1)
2100	56.2	144.4	216.6	345.0	12–18
	(22.1)	(56.9)	(85.3)	(135.8)	(4.7–7.1)

Source: From J. Hoffman, D. Keyes, and J. Titus, 1983, *Projecting Future Sea Level Rise: Methodology, Estimates to the Year 2100, and Research Needs,* 2nd rev. ed., U.S. GPO No. 055-000-00236-3, Washington, D.C.: Government Printing Office.

Note: Scenarios recorded here differ slightly from those in other chapters because of refinements made in the treatment of trace gases in the second revised edition.

[a]Called the low scenario in other chapters.

Table 3-2. Summary of Special Case Scenarios (in cm, with inches in parentheses)

Year	Minimal[a]	Maximal[b]
2000	1.1	19.5
	(0.43)	(7.7)
2025	3.3	64.8
	(1.3)	(25.5)
2050	6.5	130.7
	(2.6)	(51.5)
2075	10.9	259.2
	(4.3)	(102.0)
2100	17.0	439.0
	(6.7)	(172.8)

Source: From J. Hoffman, D. Keyes, and J. Titus, 1983, *Projecting Future Sea Level Rise: Methodology, Estimates to the Year 2100, and Research Needs,* 2nd rev. ed., U.S. GPO No. 055-000-00236-3, Washington, D.C.: Govrnment Printing Office.

Note: Estimates shown here differ from those in the second revised edition of Hoffmann et al.

[a]Declining solar, decreasing volcanic, chlorofluorocarbon (CFC) emissions capped at 1980 emission levels, 0.1 percent rise in N_2O, 0.5 percent increase in methane, 1.5°C rise, no ice contribution.

[b]4.5°C thermal equilibrium, increasing solar; 1.9 cm^2 sec^{-1} diffusion; CFC grows 4.5 percent of 1980 level, N_2O at 0.9 percent per year, methane at 2.5 percent per year, 2:1 ice discharge to thermal expansion ratio.

research will be needed. Merely waiting for observations will, however, be the slowest way to learn more about sea level's future rise. To maximize the value of future observations, the theoretical base and models used to interpret the relevant data must be improved. Rapid progress can be made by accelerating the research aimed at improving our basic understanding of the processes that underlie climatic change and sea level rise.

Unfortunately, a serious acceleration of research will require additional resources. The present shortage of federal funds has already reduced research in many of the areas of greatest concern. Therefore, three demonstrations need to be made to justify changing this situation and accelerating research.

First, a demonstration of the value to society of speeding the development of better information must be made. This need is documented in other chapters and will not be discussed here. Second, the possibility of speeding research to narrow the range of sea level rise and the probable progress under different funding levels must be demonstrated. Finally, it must be demonstrated that the appropriate organizational and management processes can be used to ensure that research is effective and accomplishes what is theoretically possible. Since these last two areas are closely linked to the scenario-generating process itself, they will be given a brief review here.

Opportunities for Narrowing the Range of Estimates

The range of scenarios can be narrowed through a series of short- and long-term projects. There are a number of possibilities for short-term projects during the next two years that utilize existing knowledge to make better projections.

In the area of atmospheric composition, existing information on trace gases could be accumulated and used to generate more realistic scenarios. Parametric models that allowed sensitivity testing could be used that join sinks, sources, and exchanges to generate better low and high scenarios for these gases. Carbon cycle models could also be parametrically extended to look at major uncertainties in ocean uptake and photosynthesis.

For thermal responses, transient (year-by-year) runs could be made using general circulation models (GCMs). This would allow better estimates of temperature increases and thermal fluxes through time on a geographically disaggregated basis. Net ablation due to melting, evaporation, sublimation, and additional snowfall in polar regions could be tracked in these runs.

Scenarios of the deglaciation of icefields could be constructed using models that represent physical processes. Critical parameters, relationships, and boundary conditions would be varied, thus generating a first good estimate of the plausible ranges of deglaciation contribution.

Together, these efforts would greatly increase the total confidence in the scenarios, although they might not narrow the range very much. (Whether they would or not depends almost totally on the output of the glacial research.)

In the longer term, improving scenarios and narrowing the estimates of future sea level rise must be based on research that produces greater knowledge, observations, and modeling capabilities useful for simulating the future evolution of the relevant natural systems. Opportunities for speeding the development of knowledge exist for all systems that determine sea level's future rise.

Most critically, more research needs to be conducted on trace gases, including monitoring and modeling. Current funding efforts in this area are relatively small and dispersed through many federal agencies. Better observational systems and better models of atmospheric composition can be developed and refined, greatly increasing our confidence in predictions of future trace gas concentrations.

For climate response, several areas of improvement can be targeted. First, and of paramount importance, dynamic ocean models must be integrated into general circulation models (GCMs). This task will require a much larger effort than the resources currently devoted to this task. Yet until this is done, it will be impossible to have an accurate understanding of the geographic distributions of future precipitation and temperature changes that will be critical to better projections of snow accumulation and melting. Second, cloud responses need to be modeled better in GCMs and validated with observational data. Such efforts would appreciably reduce uncertainty about the thermal equilibrium. Third, larger computers and more computer time should be provided to run general circulation models. At present, the number of runs (experiments), their geographical scale, and the representation of processes in models are severely limited by lack of computational support. Last, albedo, sea ice, and hydrological processes need to be modeled more carefully in large climate models. Improvements in these areas would improve the accuracy in projecting global warming, and provide information on the accumulation/melting of snow and ice and on the conditions critical to projecting deglaciation.

Of greatest importance to improving sea level rise estimates, however, will be development of a better understanding of snow/ice responses. Three primary areas regarding ocean-glacial response can be improved. First, observational programs to track the mass balance of all icefields and sea level throughout the world could be undertaken to refine the basis for validating models. Second, for each icefield, models could be developed that can realistically consider changes in conditions predicted by climate models, as well as the actual geography of fields. Field work will be needed to provide data to these efforts. This would yield much better estimates of deglaciation. Last, selected experiments could be conducted, such as pulling icebergs to waters of appropriate temperature, to learn the true value of certain critical parameters.

These lists of research opportunities represent a partial description of possibilities to accelerate the acquisition of knowledge of sea level rise. Existing funding by various federal agencies "supports" the accomplishment of all these goals to a very limited degree. Unfortunately, many projects that are critical to the timely accomplishment of these goals are not being funded, or are being funded at insufficient levels, or with inadequate guarantees of long-term support. For example, many opportunities to observe natural systems and collect data are being lost, disrupting or delaying the construction of time series and geographical data sets that are critical to improving or validating models. And no major effort to model oceans and incorporate these models into climate models has been undertaken. Present funding levels could significantly delay the date at which more precise and reliable estimates of sea level rise are available.

Management of Research

The research undertaken to improve estimates of sea level rise must consider the need for secure, long-term commitments of funding. Major interdisciplinary scientific teams, not just individual or group projects, will be needed to address most of the scientific issues that are blocking more precise estimates of sea level rise. The development and maturation of research efforts depends on steadily increasing the funding of teams that are directed by well-respected scientists capable of integrating the efforts of forcefully independent scientists.

Success in scientific research can never be guaranteed—those at the frontier cannot necessarily predict what is beyond or how fast they will be able to proceed. Success can be thwarted, however, by failing to sustain the conditions needed for performing solid basic research. In the case of sea level rise, the challenges to progress are great, in part, because of the interdisciplinary nature of the efforts required. The opportunities for progress that exist can be successfully pursued only if interested parties decide that the value of accelerating research justifies the costs. The question is not whether we can do better, but whether we have the will to do so.

NOTES

1. J. Charney, chairman, Climate Research Board, 1979, *Carbon Dioxide and Climate: A Scientific Assessment*, Washington, D.C.: NAS Press. This panel made a major review of the evidence. After looking for factors that might diminish the warming to negligible proportions, the panel concluded that a significant rise in temperature was almost certain. A second panel reviewed the work done since the initial assessment and concurred with its results: J. Smagorinsky, chairman, Climate Research Board, 1983, *Carbon Dioxide: A Second Assessment*, Washington, D.C.: NAS Press.
2. The estimates of scenarios used in this chapter are based upon results contained in J. Hoffman et al. (1983). They differ from the scenarios used in other chapters of this book in having a slightly lower high scenario, a higher conservative scenario, and higher mid-range scenarios. These estimates, based on later computer runs using more realistic estimates of trace gas growth, should be used for future analyses until they are improved upon by the next generation of scenarios, due to be published by EPA in winter 1985.
3. Estimates made by Lovins et al. (1981) provide an interesting counterexample in which radically slower energy use rates are assumed. The feasibility of these estimates is in doubt, however.
4. The choice of these growth rates depends on the beliefs one has about sinks and sources. Sinks can become saturated and sources exhausted. New sources or sinks can develop with a climate change. Any projection of future levels of trace gases implies some combination of changes in sources and sinks. To the degree that reality is different, the projections will be wrong. For a discussion of these gases, see Chamberlain, Joseph W., Henry M. Foley, Gordon J. MacDonald, and Marvin A. Ruderman. 1982. "Climatic Effects of Minor Atmospheric Constituents." In W. Clark, ed., *Carbon Dioxide Review*. New York: Clarendon Press, pp. 253-277.
5. While the NAS has estimated between $1.5°C$ and $4.5°C$ as the equilibrium response to doubled CO_2, no major modeling effort has yielded an estimate below $2°C$. Recently, results from the Goddard Institute for Space Studies and the National Center for Atmospheric Research have yielded results of around $4°C$ (pers. comm.).
6. Personal communication with Dr. Gary Russell on his estimate of thermal expansion with heat fluxes made for multiple columns at different temperatures.

REFERENCES

Bacastow, R., and D. Keeling. 1979. "Models to Predict Future Atmospheric CO_2." In W. Elliot and L. Machta, eds., *Workshop on the Global Effects of Carbon Dioxide from Fossil Fuels*. Washington, D.C.: Department of Energy, pp. 72-90.

Bach, W. 1981. "Fossil Fuel Resources and Their Impacts on Environment and Climate." *International Journal of Hydrogen Energy* 6:185-201.

Barnett, T. P. 1983. "Global Sea-Level: Estimating and Explaining Apparent Changes." In O. T. Magoon, ed., *Coastal Zone '83*. New York: American Society of Civil Engineers, pp. 2777-2784.

Barnola, J. M., D. Raynaud, H. Oeschger, and A. Neftel. 1983. "Comparison of CO_2 Measurements by Two Laboratories on Air from Bubbles in Polar Ice." *Nature* **303**:410.

Bentley, C. 1983. "The West Antarctic Ice Sheet: Diagnosis and Prognosis." In *Proceedings: Carbon Dioxide Research Conference: Carbon Dioxide, Science, and Consensus.* Conference 820970. Washington, D.C.: Department of Energy, pp. IV.3-IV.50.

Boesch, D. F., ed. 1982. *Proceedings of the Conference on Coastal Erosion and Wetland Modification in Louisiana: Causes, Consequences, and Options.* FWS-OBS-82/59. Washington, D.C.: Fish and Wildlife Service, Biological Services Program.

Broecker, W. S., T-H. Peng, and R. Engh. 1980. "Modeling the Carbon System." *Radiocarbon* **22**:565-598.

Charney, J., chairman, Climate Research Board. 1979. *Carbon Dioxide and Climate: A Scientific Assessment.* Washington, D.C.: NAS Press.

Clark, J. A., W. E. Farrell, and W. R. Peltier. 1978. "Global Changes in Post Glacial Sea Level: A Numerical Calculation." *Quaternary Research* **9**:265-287.

Edmonds, J., and J. Reilly. 1981. *Determinants of Energy Supply.* Oak Ridge, Tenn.: Oak Ridge Associated Universities, Institute for Energy Analysis.

Edmonds, J., and J. Reilly. 1982. *Global Energy and CO_2 to the Year 2050.* Oak Ridge, Tenn.: Oak Ridge Associated Universities, Institute for Energy Analysis.

Gibbs, Michael. 1983. *Current and Future CFC Production and Use.* Washington, D.C.: ICF Incorporated.

Gornitz, V., S. Lebedeff, and J. Hansen. 1982. "Global Sea Level Trend in the Past Century." *Science* **215**:1611-1614.

Haefle, W. 1981. *Energy in a Finite World.* Cambridge, Mass. Ballinger.

Hansen, J. E., D. Johnson, A. Lacis, S. Lebedeff, D. Rind, and G. Russell. 1981. "Climate Impact of Increasing Atmospheric Carbon Dioxide." *Science* **213**:957-966.

Hansen, J. E., A. Lacis, and D. Rind. 1983. "Climate Trends Due to Increasing Greenhouse Gases." In O. T. Magoon, ed., *Coastal Zone '83.* New York: American Society of Civil Engineers, pp. 2796-2810.

Hansen, J. E., W. C. Wang, and A. N. Lacis. 1978. "Mount Agung Eruption Provides Test of Global Climatic Perturbation." *Science* **199**:1065-1068.

Hays, J. D., J. Imbris, and N. J. Shackelton. 1976. "Variations in the Earth's Orbit, Pacemaker of the Ice Ages." *Science* **194**:1121-1132.

Hoffman, J., D. Keyes, and J. Titus. 1983. *Projecting Future Sea Level Rise: Methodology, Estimates to the Year 2100, and Research Needs.* 2nd rev. ed. U.S. GPO No. 055-000-00236-3. Washington, D.C.: Government Printing Office.

Hughes, T. 1983. "The Stability of the West Antarctic Ice Sheet: What Has Happened and What Will Happen." In *Proceedings: Carbon Dioxide Research Conference: Carbon Dioxide, Science, and Consensus.* DOE Conference 820970. Washington, D.C.: Department of Energy, pp. IV.51-IV.73.

Keeling, C. D., R. B. Bacastow, and T. P. Whorf. 1976."Atmospheric Carbon Dioxide Variations at Mauna Loa Observatory, Hawaii." *Tellus* **28**:28.

Keyfitz, N., E. Allen, J. Edmonds, R. Dougher, and B. Widget. 1983. *Global Population, 1975-2075, and Labor Force 1975-2050.* Oak Ridge, Tenn.: Oak Ridge Associated Universities, Institute for Energy Analysis.

Keyes, D., J. Hoffman, S. Seidel, and J. Titus. 1984. "EPA's Studies of the Greenhouse Effect," *Science* 223:538-539.

Killough, G., and W. Emmanuel. 1981. *A Comparison of Several Models of Carbon Turnover in the Ocean with Respect to Their Distributions of Transit Time and Age and Responses to Atmospheric CO_2 and "C."* Publication 1602. Oak Ridge, Tenn.: Oak Ridge National Laboratories, Environmental Science Division.

Lacis, A., J. E. Hanson, P. Lee, T. Mitchell, and S. Lebedeff. 1981. "Greenhouse Effect of Trace Gases, 1970-1980." *Geophysical Research Letters* 8(10):1035-1038.

Lemon, E., ed. 1983. *CO_2 and Plants: The Response of Plants to Rising Levels of Atmospheric Carbon Dioxide.* Boulder: Westview Press.

Lovins, A., L. Lovins, F. Krause, and W. Bach. 1981. *Least Cost Energy.* Andover, Mass.: Brick House Publishing Company.

Lugo, Ariel. 1980. "Are Tropical Forest Ecosystems a New Sink?" In S. Brown, A. Lugo, and B. Liegel, eds., *The Role of Tropical Forests in the World Carbon Cycle.* Washington, D.C.: Department of Energy, pp. 1-18.

Manabe, S. 1983. "Carbon Dioxide and Climatic Change." *Advances in Geophysics.* New York: Academic Press.

McDonald, G., ed. 1982. *The Long-Term Impacts of Increasing Atmospheric Carbon Dioxide Levels.* Cambridge, Mass.: Ballinger.

Mercer, J. H. 1978. "West Antarctic Ice Sheet and CO_2 Greenhouse Effect: A Threat of Disaster." *Nature* 271(5643):321-325.

Rasmussen, R. A., and M. A. K. Khalil. 1981. "Increase in the Concentration of Atmospheric Methane." *Atmospheric Environment* 15(5):883-886.

Rotty, R. M. 1983. "Distribution of Changes in Industrial Carbon Dioxide Production." *Journal of Geophysical Research* 88:1301.

Seidel, Stephen and Dale Keyes, Environmental Protection Agency. 1983. *Can We Delay a Greenhouse Warming?* Washington, D.C.: Government Printing Office.

Weiss, R. F., Keeling, C. D., and H. Craig. 1981. "The Determination of Tropospheric Nitrous Oxide." *Journal of Geophysical Research* 86(68):7197-7202.

Wilson, R. C. 1982. "Solar Irradiance Variations and Solar Activity." *Journal of Geophysical Research* 87:4319-4326.

Woodwell, G. 1978. "The Carbon Dioxide Question." *Scientific American* 238:234-243.

Chapter 4

The Physical Impact of Sea Level Rise in the Area of Charleston, South Carolina

Timothy W. Kana, Jacqueline Michel,
Miles O. Hayes, and John R. Jensen

INTRODUCTION

This chapter reports on a pilot study to determine the shoreline impact from accelerated rises in sea level due to anthropogenic (man-induced) factors. The methods developed have been applied to the coastal city of Charleston, South Carolina, to determine the effects of various accelerated sea level rise scenarios for the years 2025 and 2075.

In the last few decades, there have been numerous studies on the trends and rates of both eustatic and local sea level changes. Eustatic changes are global in nature due to a general rise of the sea level compared to local changes for a specific area due to the relative rise or subsidence of the land surface with respect to a stationary, general sea level. There has been an overall rise in sea level of about 40 m (130 ft) since the last glacial epoch, called the Wisconsin ice age, which ended about 14,000 years ago. From 7,000 to 3,000 years ago, sea level along the east coast of the United States rose at a rate of about 0.3 cm (0.1 in) per year (Kraft, 1971). Studies of sea level over the last two centuries have estimated that global sea level is rising at a rate of 0.10-0.12 cm/yr (0.04-0.05 in/yr). For the Charleston case study area, Hicks and others (1978, 1983) have estimated

Table 4-1. Sea Level Rise Scenarios for the Charleston Case Study Area (in cm, ft in parentheses)

	Year		
Scenario[a]	*1980*	*2025*	*2075*
Baseline	0	11.2 (0.4)	23.8 (0.8)
Low	–	28.2 (0.9)	87.0 (2.9)
Medium	–	46.0 (1.5)	159.2 (5.2)
High	–	63.8 (2.1)	231.6 (7.6)

Source: Global sea level rise scenarios are from Chapter 3, modified to reflect local conditions based on the historical trend for Charleston. (S.D. Hicks et al., 1983, *Sea Level Variations for the United States, 1855–1980,* technical report, Rockville, Md.: NOAA, Tides and Water Levels Branch.)

[a]Baseline scenarios for each year reflect present trends. Other scenarios reflect accelerated sea level rises at various rates.

that the total sea level rise since 1922 has been 0.25 cm/yr (0.1 in/yr).* For our analysis, the local rate was assumed to be 0.25 cm/yr (0.10 in/yr), and the eustatic rates used were a baseline of 0.12 cm/yr (0.05 in/yr) and the low, medium, and high scenarios discussed in Chapters 1 and 3. These scenarios are outlined in Table 4-1 for the years 2025 and 2075.

This chapter describes the physical responses of coastal land forms in Charleston to accelerated sea level rise. Three types of response are addressed: shoreline changes due to landward displacement of the water line after a sea level rise (in some geomorphic settings, where sediment supply is great, the shoreline may accrete or keep pace with a sea level rise.); storm surges that affect new or higher elevations after a sea level rise; and groundwater changes caused by the intrusion of seawater to higher levels in aquifers.

The chapter is organized as follows. First, the Charleston case study area is described. Then, in turn, we discuss the methodology used in the study: modeling shoreline changes, mapping methods, historical shoreline trends, and storm surge and groundwater analyses. Finally, the results and an analysis of the methodology used are presented.

CHARLESTON CASE STUDY AREA

History of Human Development

The first European settlers arrived in Charleston around 1670. Since that time, the peninsula city has undergone dramatic shoreline changes,

*Based on a global (eustatic) rise of 0.12 cm/yr (0.05 in/yr) plus local subsidence of 0.13 cm/yr (0.05 in/yr).

predominantly by landfilling of the intertidal zone. Early maps show that over one-third of the peninsula has been "reclaimed." Much of the landfilling occurred on the southern tip of Charleston, behind a high seawall and promenade, known as the Battery. Many of the buildings on the lower peninsula are of historic value and play an important role in the area's major industry—tourism. These areas already experience frequent flooding during intense rainstorms and unusually high tides and would have high priority for any protection/mitigation actions to prevent further flooding due to sea level rise.

The port of Charleston, which dominates the eastern shore of the city, has an active merchant ship port, along with a large U.S. Navy base along the Cooper River (Figure 4-1, the area described is in the vicinity of station number 29). Maintenance of the ship channels to the port has generated large volumes of dredge spoil, which have been disposed of at every possible nearby site. There are only two sites currently authorized

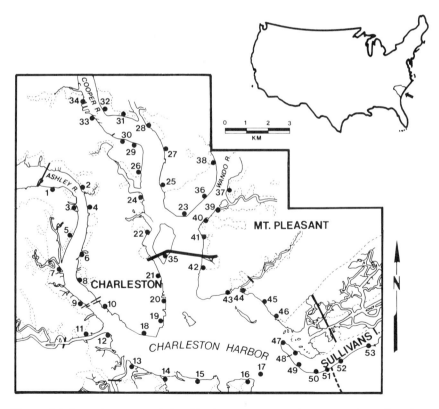

Figure 4-1. Location map of the Charleston case study area and 53 shoreline stations used in the historical trend analysis.

for spoil disposal, and the addition of other sites is unlikely. Plans call for construction of dikes as high as necessary to retain spoil in the designated sites.

The mainland to the east and west of Charleston is primarily residential; much of it is of low density. The trend has been toward slow encroachment on farmland with more intensive development near the harbor, along the Intracoastal Waterway, or on the larger creeks. Sullivans Island and Isle of Palms, developed before World War II, have a large year-round population. These barrier islands northeast of Charleston Harbor are also the principal recreational beaches for the metropolitan area.

Site Description

The Charleston area has a complex coastal plain morphology which has been significantly altered by man in the last 100 years (Figure 4-1). The outer shore to the north is composed of geologically young, developed barrier islands (e.g., Sullivans Island) which are relatively flat; elevations typically average less than 3 m (10 ft) above mean sea level (MSL) on the islands in the study area. Sheltered by the barrier islands is an extensive, intertidal salt marsh/tidal creek system. At the edge of the marsh/mainland contact (Figure 4-1, dashed line beginning at station number 46 in Mount Pleasant), there is a break in slope and a distinct change to terrestrial vegetation. Elevations on the lower Charleston peninsula are generally 3 m (10 ft), with small areas up to 5.5 m (18 ft). The study area west of theAshley River is very flat, with elevations generally about 3 m (10 ft). The Charleston shoreline has a characteristic dendritic drainage pattern typical of drowned coastal plain areas.

The highly populated Charleston peninsula is formed by the junction of three rivers which discharge into Charleston Harbor: the Cooper, Ashley, and Wando Rivers (shown in Figure 4-1). The Cooper River dominates the discharge into the harbor, with an average flow of 450 m^3/s [15,600 ft^3/s (cfs)], which includes flow from the Santee River (a large river originating in the mountains) diverted for hydroelectric power in 1942. The diversion has reportedly caused a significant increase in sedimentation in Charleston Harbor, requiring increased dredging from 400,000 m^3 (525,000 yd^3) per year to over 7,500,000 m^3 (10,000,000 yd^3) per year (S.C. Water Resources Commission, 1979). Studies have shown that diversion is responsible for 85 percent of the sedimentation in Charleston Harbor (U.S. Army Corps of Engineers, 1966). To alleviate this problem, the flow will be redirected back to the Santee River by 1985, reducing discharge to one-fifth its present volume. The natural harbor shoreline is dominated by fringing salt marsh from several meters to over a kilometer wide. As will be shown from the historical shoreline trend data, most of the marshes have accreted since diversion of the Santee into Charleston Harbor.

The entrance to Charleston Harbor has also been modified by the construction of jetties in the 1890s to stabilize the navigation channel. The jetties have caused large-scale changes in sediment transport patterns, producing up to 300 m (1,000 ft) of deposition along the barrier islands (Sullivans and Isle of Palms) to the north. Concomitant with accretion north of the harbor, extensive erosion has occurred south of the jetties, including over 500 m (1,700 ft) of erosion along Morris Island (Stephen et al., 1975). Another man-made change in the system is the Intracoastal Waterway, dredged to 4 m (12 ft), which has altered flow patterns in the marsh behind the barrier islands.

Physical Processes

South Carolina's climate is mild, with an average temperature for the coastal region ranging between 10.1°C (50.2°F) in December and 27.2°C (81.0°F) in July. An average of 1.4 hurricanes and tropical storms affect the coast annually. Winds are somewhat seasonal, with northerly components dominating in fall and winter and southerly components dominating in spring and summer (Landers, 1970).

The tidal range increases considerably from north to south along the state's shoreline, from approximately 1.7 m (5.5 ft) at the northern border to 2.7 m (8.8 ft) at the southern border. The increasing tidal prism (volume of water flowing in and out of a harbor or estuary with the movement of the tides) has several effects as one moves southward along the South Carolina coast: tidal inlets become more frequent and are larger in order to accommodate greater tidal flow, salt marshes are more extensive, and the ebb-tidal deltas (seaward shoals at inlets) become much larger (Nummedal et al., 1977). Charleston's mean tidal range is 1.6 m (5.2 ft); spring tides average 1.9 m (6.1 ft); and the highest astronomic tides of the year exceed 2.1 m (7.0 ft) (U.S. Department of Commerce, 1981). The spring tidal elevation represents the limit of human development because the land surface is inundated every 14 days to that elevation, and it is the upper limit of high marsh vegetation on which development or any alteration is strictly regulated by South Carolina Coastal Zone Management laws (U.S. Department of Commerce, 1979).

The wave climate at Charleston is dependent on offshore swell conditions but is diurnally modified by the seabreeze/landbreeze cycle typically occurring in the area. The prevailing winds are from the south and west in these latitudes, but the dominant wind affecting the coastline is from the northeast, originating in extratropical storms travelling parallel to the coast (Finley, 1976). Breaking wave heights along the outer beaches average approximately 60 cm (2 ft) high in the Charleston area. Predominant wave-energy flux is directed south along the beaches, accounting for net longshore transport rates of approximately 100,000 m^3/yr (135,000 yd^3/yr) (Kana, 1977).

The relatively large tidal range produces current velocities at all tidal entrances and creeks that often exceed 1.5 m/s (5.0 ft/s) (Finley, 1976). With three major tidal rivers within the study site, a diverse set of estuarine processes influences circulation, flushing, and sedimentation patterns in Charleston Harbor.

The subtropical climate of the southeast produces high weathering rates, which provide large fluxes of sediment to the coastal area. Suspended sediment loads, which dramatically increased in Charleston Harbor because of diversion of the Santee River, provide significant inputs to the study area and may account for growth of some marsh shorelines. Marshes accrete through the settling of fine-grained sediment on the marsh surface as cordgrass (*Spartina alterniflora*) baffles the flow adjacent to tidal creeks. Marsh sedimentation has generally been able to keep up with or exceed recent sea level rises along many areas of the eastern U.S. shoreline (Ward and Domeracki, 1979).

Hydrogeology

The water table aquifer is composed of surficial sands and clays of Pleistocene age and, in the study area, extends to 10-20 m (30-65 ft) below sea level. It is heavily used by the Mount Pleasant and Sullivans Island water districts; both have over 20 wells or well-point systems, each tapping the shallow aquifer. Although the exact position of the freshwater/saltwater interface is unknown, there have been reports of shallow wells close to shore being moved because of unsuitable water quality. The next geologic unit is the Cooper Marl, a calcareous clay, which acts as a confining layer on top of the Santee Limestone-Black Mingo aquifers. These aquifers have not been used for drinking water in the area since about 1950 because of saltwater intrusion. The present freshwater/saltwater interface in this aquifer system is thought to be near Summerville, about 40 km (25 mi) inland (Drennen Parks, 1983, South Carolina Water Resources Commission, personal communication).

The Black Creek aquifer of Late Creataceous age is an important water source. Although there is no saltwater currently in the Black Creek aquifer in the study area, the U.S. Geological Survey (USGS) has measured chloride contents of 390-534 mg/l in the lower half of the aquifer on Kiawah and Seabrook Islands, about 30 km (20 mi) to the southwest. The position of the freshwater/saltwater interface in the Black Creek offshore of Charleston is unknown. The deepest aquifer used in Charleston is the Middendorf Formation; deep wells down to 700 m (2,200 ft) have not encountered saltwater in the study area. However, on Kiawah and Seabrook Islands, freshwater (62-160 mg/l chloride) was found to 700 m (2,200 ft), and saline water (1,440 mg/l chloride) was encountered at 790 m (2,400 ft).

The main users of groundwater are the municipalities of Mount Pleasant, Sullivans Island, and Isle of Palms, which use several million gallons per day. Groundwater demand is expected to grow rapidly, as these areas are projected to experience rapid population growth. The city of Charleston uses surface water and services the peninsula and west Ashley areas. The present position of the freshwater/saltwater interface for the shallow and deep aquifers is unknown, except 30 km (20 mi) to the southwest, and the middle aquifer is already too salty to use. As water usage increases, saltwater intrusion due to overpumpage alone is predicted to be a serious problem in the future, eventually resulting in abandonment of the shallow aquifer for potable water.

MODELING EFFECTS OF SEA LEVEL RISE

Shoreline Changes

With respect to retreating or eroding shorelines, there are several different shoreline response concepts that can be used to model the resulting shoreline reconfiguration as a function of sea level rise. The simplest to quantify is the inundation concept (Figure 4-2), whereby preexisting contours above shorelines are used to project new shorelines. Here, slope is the controlling factor. Shorelines with steep slopes will experience little

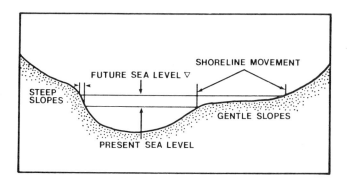

Figure 4-2. Schematic cross-section of inundation concept of sea level rise. The shoreline movement greatly depends on the land slope.

horizontal displacement of the shoreline. Gently sloping shores, on the other hand, will experience a much broader area of inundation for a given sea level rise. The inundation concept, in fact, is the preferred methodology to apply for immobile substrates or rocky or armored shorelines, or where the shoreline is not exposed to wave action or strong currents.

The analysis becomes more complicated when dealing with mobile sediments, such as sand-sized material along beaches. As Chapters 1 and 5 describe, Bruun (1962) introduced a model to predict the equilibrium adjustment of shoreline profiles during a sea level rise. Bruun hypothesized that a typical concave-upward profile in the nearshore zone will maintain its configuration, but the profile will be translated landward and upward as sediments erode near the old water level and settle in deeper water, building up the bottom. This offshore displacement of sediments theoretically maintains the same depth at a given distance from the new shoreline compared to that distance and depth combination from the old shoreline. Hands (1981) presented a relationship based on Bruun's model, which is a practical way to predict this profile adjustment:

$$y = \frac{rX}{Z} (R_A)$$

where y = shoreline change; r = change in water level; X = average, representative width of adjustment in the profile; Z = height of responding profile or vertical relief of active beach; and R_A = overfill ratio to account for loss of suspended load from the eroded material.

Nearshore surveys along Charleston's beaches (RPI, unpublished) indicate the depth of active movement in the profile (i.e., wave base) is typically at depths of 10-15 m (33-50 ft). This yields values of Z between 15 and 20 m (50-66 ft) when the mean dune elevation is added. Based on existing slopes, these values for Z yield a typical range of X between 1,000 and 3,000 m (3,300-10,000 ft) for Charleston's outer beaches. Factor R_A is 1.0 if no fine-grained suspended sediment losses are expected. We assumed this to be the case for the outer beaches since existing dune sediments essentially match the beach and nearshore sediments in the project area (Brown, 1976). Hands' model, illustrated in Figure 4-3, was tested against sandy shorelines of the Great Lakes, which responded to changes in water level. Although the formula has been shown to apply under field conditions and uses generally available information, it only applies to erodable substrates, such as sand beaches or unconsolidated bluffs.

The model for shoreline changes along beaches that we believe is presently the most realistic and feasible for widespread application combines projections of new equilibrium shorelines using historical shoreline movement patterns and the erosion/inundation effects due to sea level rise according to Hands. Once the sea level has exceeded the dune elevation, onshore movement of beach sediments occurs by washovers (Leatherman, 1977). The rate of shoreline retreat, once in the washover mode, can be estimated from retreat rates along existing washover islands north and south of the study area (Stephen et al., 1975). Thus, we project additional erosion due to sea level rise for shorelines on barrier islands. We do not project accelerated erosion along riverine (cohesive sediment) shorelines due to sea level rise.

In summary, the model for shoreline change that has been applied to Charleston consists of drowning the shoreline by each particular sea level rise scenario, then applying a shoreline correction factor for particular coastal geomorphic types that considers: historical erosion/accretion rates for beaches and active cutbanks on rivers, mobility of sediments, likelihood of the profile to respond rapidly to sea level rise and maintain its general shape, and locus of sediment movement (offshore, alongshore, or onshore) for a given site. The first factor is quantifiable, based on historical data; sediment mobility is greatest in the sand-size ranges, decreasing as sediments get coarser or very fine and cohesive. Major sediment transport patterns can be deduced from geomorphic features and man-made coastal structures.

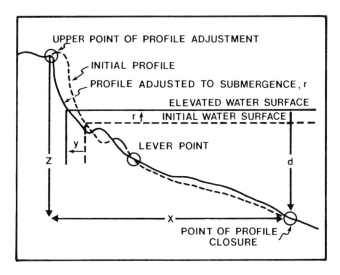

Figure 4-3. Sketch of predicted shoreline profile adjustment to a change in water elevation. *(After E.B. Hands, 1981, Predicting Adjustments in Shore and Offshore Sand Profiles on the Great Lakes, CERC technical aid 81-4, Fort Belvoir, Va.: Coastal Engineering Research Center.)*

Storm Surge

The term storm surge refers to any departure from normal water levels due to the action of storms. This can take the form of a set-up or rise in the sea surface due to excess water piling up against the shore or a set-down if water is removed from the coastal region. For obvious reasons, a super-elevation of coastal waters is of most concern because of its potential for causing property damage from flooding.

Storm surges are generally reported as a deviation in height from MSL. The magnitude of this deviation at any point along the coast is a function of several factors, including: the energy available to move excess water toward the coast (wind and waves), the width of the continental shelf, the shape of the basin, and the phase of the normal astronomic tide.

The most widely applied model for predicting open-coast hurricane-surge elevations is the National Oceanic and Atmospheric Adminstration's (NOAA) SPLASH [Special Program to List Amplitudes of Surges from Hurricanes (Jelesnianski, 1972)]. Recently, a model called SLOSH (Sea, Lake, and Overland Surges from Hurricanes), which "routes" the surge inland, has been developed by NOAA (Jelesnianski and Chen, 1984) and is considered the state of the art for inland surge computations. Unfortunately, this model was not complete for the Charleston study area at the time the study was undertaken.

Designers and engineers have set standard recurrence intervals such as 1, 10, 25, 50, or 100 years to compare flood elevations from one place to another. This can be restated as the percent chance of occurrence for a particular flood level in any year. For example, a 10-year flood elevation has a 10 percent chance of occurring each year, whereas a 100-year flood has a 1 percent chance of occurring. The relative increase in flood levels from a 10-year to a 100-year storm is generally less than 25 percent (U.S. Army Corps of Engineers, 1977). In most regions, this holds true for inland, as well as open-coast, surges. The generally accepted standard for safe design is the 100-year flood level. This is the basis for delineating flood-prone areas used by the Army Corps of Engineers, NOAA, and the Federal Insurance Administration (FIA).

Two different probability storms were used in the present study to evaluate the effect of sea level rise on flooding frequency: the 100-year storm and the threshold storm. (Threshold storm is that storm with the greatest probability of initiating significant damage in the study area.) The 100-year storm elevations ranged from 4.2 m (14 ft) on the outer beaches to 2.7 m (9 ft) inland. For Charleston, the threshold storm was selected to be the 10-year storm. It was determined by sequentially raising water levels until significant inundation of developed areas occurred. The 10-year storm elevations ranged from 2.1 m (7 ft) on the outer beaches to 1.4 m (4.5 ft) inland. Intermediate storm-surges can be selected from frequency curves on the historical tidal-storm elevations for Charleston (Myers, 1975).

Groundwater Analyses

Saltwater intrusion is the most common and serious pollutant of fresh groundwater in coastal aquifers. Although many complex mathematical models have been developed to predict saltwater intrusion, a simple concept, the Ghyben-Herzberg principle (Herzberg, 1961), can be used as a conservative estimate of the position of and change in the freshwater/ saltwater boundary.

The Ghyben-Herzberg principle predicts that the depth of the freshwater/saltwater interface is 40 times the elevation of the water table above MSL. Therefore, if the water table is 1 m above MSL, the freshwater/saltwater interface is predicted to be at 40 m below MSL at that point. For artesian aquifers (aquifers which are confined by overlying, relatively impermeable beds), the freshwater/saltwater interface can be predicted by using the elevation of the piezometric surface, which is the artesian pressure or level of water in the aquifer analogous to the water level in unconfined aquifers. A later section on results includes an explanation of our assumptions regarding the modeling of groundwater impacts from sea level rise.

MAPPING METHODS

The first step required in the analysis was to establish a method for contouring new shoreline positions and storm surge elevations for each sea level rise scenario. New positions and elevations could be plotted by manual interpolation between closest contours on standard USGS topographic maps. This procedure is appropriate for simple shorelines or small geographic areas. However, for the Charleston case study, an automated interpolation scheme was necessary for two reasons: first, the 5 ft contour interval on the existing topographic maps did not provide the necessary detail for accurate interpolation, especially between 0 and 5 ft; and second, there were well over 800 km (500 mi) of shoreline to interpolate.

Topographic maps were made by the translation of map contours using a digital map data base. Computer-generated maps were produced from digital terrain data (point elevations located on a geographical coordinate system). The maps consisted of interpolated contours generated by numerical averaging within grid squares. For example, the most accurate map would be one that has digital data plotted every few meters so that contour plotting interpolation would take place over a very small grid cell. Unfortunately, few surveys ever contain "field" data this closely spaced. Also, for practical reasons, grid spacings of a few meters would be inappropriate for a geographical area such as Charleston, which covers over 20 km^2 (75 mi^2). Instead, a compromise grid-cell spacing was required that was appropriate to the scale of the map and concentration of original contour data.

Programs using a digital terrain model (DTM) are limited to mapping with grids that fit within a designated number of rows and columns on the computer matrix. For example, if the largest matrix for a particular system is 500 rows by 500 columns, map resolution will be proportional to the scale of the map. Each grid unit on a 500 × 500 km map would represent one km^2, whereas one unit on a 500 × 500 m map could represent one m^2. The system used in the present study allowed for a 240 × 256 matrix with a grid cell for the case area of 30 m^2 (375 ft^2). This translates to map dimensions of 7.31 × 7.79 km (4.54 × 4.84 mi). The study area was approximately 3.2 times these dimensions.

Base Maps

Two types of source map were used to extract topographic/bathymetric control points. First, control points were selected from the USGS 7.5 minute topographic maps at a scale of 1:24,000 with 1.5 m (5 ft) contour intervals. Control points from this source were measured to the nearest foot. The control points were obtained by periodically sampling the contour lines and using existing benchmarks. All contours and benchmarks from −1.8 m (−6 ft) MSL up to +5.8 m (+19 ft) MSL were sampled.

An additional map source covering the city of Charleston (1:2,400 planimetric maps with 1 ft contour intervals) was used to supplement the digital topography data. Only benchmark data (no contours) were used in this data set. Control points from the large-scale maps were digitized at a resolution of 0.03 m (0.1 ft), substantially improving the quality of the DTM-computer-generated map, compared with using only data from the 1:24,000 scale USGS quadrangles. This procedure is recommended wherever additional, more accurate map sources are available.

Digitization

The spatial resolution of the DTM was chosen to be 30 m (100 ft) on the 1:24,000 base map. The elevation matrices for the study area were generated with dimensions of 240 rows by 256 columns [7.31 × 7.79 km (4.54 × 4.84 mi)]. A total of 3.5 maps was required (2,000-2,500 data points each) to cover the entire project area. A two-phase interpolation algorithm was employed to estimate the elevation values for all 900-m^2 (9,700-ft^2) cells. The first phase performed a quadrant search around each cell in question to ensure that control points would be obtained from at least two of the compass directions. A nearest-neighbor method then automatically selected, from the subset of control points generated initially, the n nearest neighbors to estimate the elevation of each cell. The interpolation was to the nearest 0.03 m (0.1 ft), resulting in a DTM with relatively accurate elevation data.

Contour maps were generated and overlaid onto the 1:24,000 base map to determine the planimetric and topographic accuracy of the interpolated grid matrix. When discrepancies occurred, additional control points were located and digitized, and a new grid matrix was created by the same interpolation method described above. This procedure was repeated several times to improve resolution as much as possible within the size limits of each grid cell.

Within the case study area, the largest sections of questionable map data are the marsh shorelines. In general, few elevation data are given on maps to illustrate the marsh topography. USGS quadrangles typically show only the MSL and 1.5 m (+5 ft) MSL contour. A computerized interpolation of intermediate elevations within the marsh would produce an unrealistic profile of the marsh surface.

During previous field surveys by our research group, it was found that a marsh has a characteristic elevation that varies with local tidal range and type of marsh vegetation (Ward and Domeracki, 1979). Figure 4-4 illustrates a typical marsh/tidal creek system for the Charleston area (a shoreline type representative of over 75 percent of the study area). Typical elevations range from +0.5 to +1.0 m (+1.5 to +3.1 ft) MSL. Note

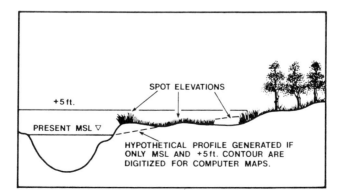

Figure 4-4. Typical South Carolina marsh transect illustrating need for spot elevations on marsh surface to improve contour interpolation.

the profile of the "typical" marsh in comparison to a hypothetical profile generated by straight interpolation between the MSL and +1.5 m (+5 ft) MSL contour. By means of aerial photographs, seaward edges of the marsh were identified and additional data points were added for the computer maps in order to account for this characteristic morphology. This gave the computer additional geomorphic data to produce more realistic shoreline interpolations.

Although efforts to add extra detail in the digital terrain model were time consuming, high concentrations of elevation data substantially improved the accuracy of the computer-generated map and allowed resolution of subtle changes in topography, a key factor for some of the smaller sea level rise scenarios. Once the digital map data base was established, the computer easily performed contour interpolation for any specified elevation. The system used is capable of plotting contour maps showing only those contours of interest. It also can display color maps on a high resolution raster CRT (Cathode Ray Tube monitor), which allowed easier visualization of the effect of sea level changes.

Computer-Generated Maps

Contour/bathymetric maps displaying the desired contours and contour intervals were prepared, scaled to overlay the original 1:24,000 scale base map. Various combinations of contours and contour intervals were plotted, depending upon the sea level rise scenario selected. These computer-generated maps became the new base maps for final determination of shoreline position using geomorphic data and increased storm surge elevations. Vertical resolution of contours was to the nearest 0.03 m (0.1 ft), whereas spatial resolution was ±15m (50 ft).

The color CRT allowed viewing various sea level rise scenarios applying the simple inundation concept. By choosing colors illustrative of water, intertidal, and land areas, it was possible to obtain a preliminary picture of the effect of each sea level scenario. The digital terrain elevation values were converted to 8 bit (byte) data ranging from values of 0 to 255. Selected elevation class intervals were assigned different colors to represent baseline and predicted changes in sea level and storm surge elevations. Although the CRT screen does not offer permanent hard copy for detailed analysis, it can be photographed directly for illustrative purposes. This is one of the most useful modern tools for applications of this kind.

HISTORICAL SHORELINE TRENDS

The computer-generated contour maps were used to project the shoreline position due to simple inundation by each sea level rise elevation. The next step was to adjust shoreline positions based on geo-morphic factors, such as historical trends of erosion and accretion, and accelerated erosion of the beach shorelines due to accelerated sea level rise, applying Hands' (1981) model.

Shorelines composed of mobile sediments, such as the beaches along the case study area, change in response to many factors. Storms, hurri-canes, and sand bypassing at inlets can cause short-term erosional and depositional trends along the shore. Long-term trends result from changes in sediment supply (such as damming or diversion of rivers) and sea level. An analysis of the net effect of accelerated sea level rise on shoreline position must exclude existing erosional/depositional trends, including those due to recent sea level rise. To accomplish this, "baseline" maps for the years 2025 and 2075 were produced that represent the predicted shoreline position at that time without any effects from accelerated sea level rise.

The baseline maps were constructed through an analysis of historical shoreline trends using aerial photographs and topographic maps available for the period 1939-1981. A total of 53 selected reference points, identifiable on successive photographs or maps, were established throughout the Charleston study area (see Figure 4-1). The distance from the reference point to the shoreline was measured on each available photograph, making the necessary scale corrections between photo sets taken at different altitudes. The trends in changes between successive photographs were used to evaluate the validity of the net change and excursion rate (shoreline movement per year) for each reference point for the years 1939-1981. Table 4-2 lists the shoreline change rates determined for each station. Annualized excursion rates were then projected into the future to compute the position of the shoreline for the reference years 2025 and 2075. The computed position was finally adjusted considering several factors:

Table 4-2. General Description of Stations and Historical Trends

Station Number	Historical Trend ft/yr (yrs of record)[a]	Geomorphic Type	Slope (MSL±5ft)
		Ashley River	
1	+ 1.6 (40)	Marsh	.001
2	− 3.4 (42)	Cutbank	.06
3	+19.4 (42)	Marsh/Point Bar	.0004
4	− 2.3 (42)	Cutbank	.032
5	+ 0.9 (34)	Exposed Marsh	.0006
6	+ 6.4 (34)	Marsh/Tidal Flat	.006
7	− 0.1 (34)	Marsh/Cutbank	.001
8	+11.1 (34)	Marsh/Point Bar	.001

[a]Accretion = +; erosion = −.

(continued)

Table 4-2. (continued)

Station Number	Historical Trend ft/yr (yrs of record)[a]	Geomorphic Type	Slope (MSL±5ft)
9	+17.5 (34)	Marsh	.002
10	+15.8 (34)	Marina	.083
		Charleston Harbor	
11	+13.8 (34)	Marsh	.005
12	+ 8.8 (34)	Marsh	.0005
13	− 3.9 (42)	Exposed Marsh	.001
14	− 0.3 (42)	Exposed Marsh	.002
15	+ 2.3 (42)	Exposed Marsh	.048
16	+37.5 (8)	Marsh/Tidal Flat	.001
17	0 (42)	Armored	Vertical
18	0 (42)	Armored	Vertical
19	+ 2.6 (34)	Armored	Vertical
20	0 (42)	Armored	Vertical
21	0 (42)	Armored	Vertical
		Cooper River	
22	+ 3.8 (42)	Marsh/Spoil	.008
23	0 (42)	Spoil Dike	.24
24	+16.0 (42)	Marsh	.004
25	+40.2 (4)	Marsh/Spoil Island	.008
26	+ 4.8 (42)	Marsh/Spoil	.006
27	+28.8 (4)	Marsh/Spoil	.009
28	+ 1.4 (4)	Sandy Marsh	.048
29	0 (42)	Armor/Bulkhead	Vertical
30	0 (42)	Armor/Riprap	.048
31	+ 6.1 (38)	Marsh	.007
32	+ 1.5 (38)	Marsh/Spoil	.012
33	0 (42)	Armor/Bulkhead	Vertical
34	0 (42)	Armor/Bulkhead	Vertical
		Wando River	
35	+ 2.0 (42)	Marsh/Spoil/Flat	.032
36	0 (4)	Spoil Dike	.24
37	0 (42)	Armor	Vertical
38	+ 0.5 (4)	Marsh	.003
39	+ 3.0 (10)	Armor/Fringe Marsh	.048
40	+ 3.0 (10)	Fringing Marsh	.016
41	+24.5 (4)	Fringing Marsh	.003
		Charleston Harbor	
42	− 5.1 (42)	Exposed Marsh	.003
43	+19.3 (42)	Exposed Marsh/Spoil	.007
44	+ 8.7 (42)	Marsh Spit	.048
45	+ 3.4 (40)	Fringing Marsh	.01
46	+ 3.4 (40)	Fringing Marsh	.012

[a]Accretion = +; erosion = −.

(continued)

Table 4-2. *(continued)*

Station Number	Historical Trend ft/yr (yrs of record)[a]	Geomorphic Type	Slope (MSL±5ft)
		Sullivans Island	
47	− 0.3 (40)	Armor/Wall	Vertical
48	− 0.2 (40)	Fringing Marsh	.008
49	− 0.5 (34)	Pocket Beach/Channel	.029
50	+21.1 (40)	Beach/Recurved Spit	.015
51	+ 1.6 (34)	Beach/Recurved Spit	0.17
52	+ 1.0 (34)	Beach/Recurved Spit	.012
53	+ 5.7 (34)	Barrier Beach	.02

[a]Accretion = +; erosion = −.

nearshore slopes, proximity to channels (if accreting), proximity to highland (if eroding);

types of sediment (e.g., cohesive marsh clays compared to unconsolidated sand deposits);

proximity to open fetches or commercial waterways;

dredge and fill (i.e., artificial changes) during the period of record, if known;

presence of unprotected development that likely would not be allowed to erode past a certain point, at which time armoring would be placed along the shoreline;

large-scale changes in sediment input that are expected to occur during the interval under consideration, such as rediversion of the Santee River.

Discrete shoreline data points were used as the basis for interpolating continuous contours for each baseline map. Because these maps were based on historical trends, they inherently include effects from recent changes in sea level. Figure 4-5 shows the 50-year trend in sea level along the Charleston shoreline with respect to adjacent land, based on tidal data for selected east coast cities (Hicks and Crosby, 1974). The sea level rise scenarios used in this study range from 2.5 to 10 times the previous rates for Charleston.

Treatment of Man-Made Shorelines

The geomorphic approach to determining historical shoreline trends is inappropriate to certain developed or man-made shorelines. Within historical times, man has manipulated shorelines to suit requirements for waterborne commerce and port development. The city of Charleston has been an active port for over 200 years and contains numerous waterfront

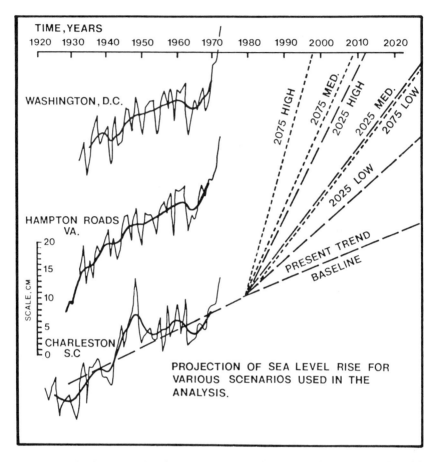

Figure 4-5. Change in sea level with respect to adjacent land for stations from the District of Columbia to South Carolina. The projected changes of the accelerated sea-level rises for 2025 and 2075 (low, medium, and high scenarios) are shown for comparison with baseline. *(After S. D. Hicks and J. E. Crosby, 1974, Trend and Variability of Yearly Mean Sea Level, 1893–1972, NOAA technical memorandum NOS-13, Rockville, Md.: Department of Commerce.)*

areas "armored" with seawalls, bulkheads, or riprap (a mat of stone along the bank) that preclude virtually all shoreline movement. Areas such as these will experience little or no change in shoreline position until a sea level rise or possible storm surge overtops the shoreline armoring. For the present analysis, maps for each scenario assumed that no alterations of the existing elevations of man-made structures occurred and that no storms would significantly erode backshore areas. All elevations used throughout this study are existing elevations. Thus, once sea level topped the structures, inundation of the backshore area proceeded according to the land slope.

Treatment of Marsh Shorelines

Shorelines fronted by marshes were treated differently than sand beaches because they do not maintain an equilibrium profile with sea level changes. Marsh surfaces accrete through the deposition of fine-grained, suspended sediment when water flow is baffled by marsh vegetation. Erosion of marshes is a slow process of wave erosion at the seaward edge of the marsh or at cutbanks of meandering streams. As shown in Table 4-2, most of the marsh stations in Charleston have been accretionary since 1939. However, the rediversion of the Santee River is expected to reduce the sediment input by 85 percent, and the marshes are not likely to continue accreting as in the recent past. Our analysis did not assume that marsh sedimentation would keep pace with sea level rise in the (low to high) scenarios. Therefore, a sea level rise would result in significant flooding of areas that are now marshes. Where marshes exist, there tends to be a critical elevation range for the majority of the deposit. In Charleston, that range is from +0.5 m (1.5 ft) to the highest normal level of tidal inundation, referred to as mean spring high water (MSHW). An incremental rise in MSL is expected to have less effect on the MSL shoreline position (since it generally occurs along steep, tidal creek banks) than on the position of MSHW because of the local slopes involved. MSHW is a critical elevation in Charleston because it establishes the contact between marsh and upland forests as well as the practical limit of development. Table 4-3 indicates the typical zonation of marsh/tidal flat habitats by elevation in the Charleston study area.

Table 4-3. Typical Elevations of a Marsh/Tidal Creek System in Charleston

Elevations	Zone	Species	Elevation (MSL) m (ft)
Highest	High marsh	*Spartina patens, Distichlus sp.*	+0.8–+1.2 (+2.5–+4.0)
	Low marsh	*Spartina alterniflora*	+0.3–+0.8 (+1.0–+2.5)
	Mud flat/oysters	*Crassostrea virginica*	−0.4–+0.3 (−1.5–+1.0)
Lowest	Channels	Benthic fauna	Less than −0.8 (Less than −2.5)

Source: Research Planning Institute, Columbia, S.C., unpublished survey data.

The response of marshes to rapid sea level rise would be by inundation, shift in vegetation zones, and creation of new intertidal habitats, rather than alteration of the substrate topography. Therefore, shoreline changes along marshes were made by showing the area of inundation using the change in MSHW for each scenario. We do not anticipate that there would be any other factors causing changes in the position of marsh shorelines, even considering the larger sea level rises that would flood the fringing highland areas. Marsh vegetation is very rapidly established and will always occupy the niche between MSL and spring high tide in sheltered areas, even in sandy substrate. Marsh vegetation would shift from high marsh to low marsh with sea level rise and would produce a wide, shallow platform that would attenuate wave energy in much the same manner as existing shorelines. Studies of shoreline changes of sheltered environments of Pleistocene sea level rises have shown that there is an upward and landward shift of environments as opposed to a one-dimensional shoreline retreat (Colquhoun et al., 1972).

Determination of Shoreline Change

The shoreline position (at mean high spring tide) for each of the 53 stations in the study area was computed for all scenarios at the years 2025 and 2075, respectively. Because of the large reduction in sediment input anticipated when the Santee River is rediverted, the marshes were assumed to go into a stable phase with no change projected from the historical trends, which are accretionary. The only shoreline change in the marsh stations for the baseline maps was assumed to be by inundation due to the continued historical rise in sea level at 0.25 cm/yr (0.1 in/yr). The total baseline change in the position of sandy shorelines (station numbers 49-53) for each scenario year included both extrapolation of historical trends (in ft/yr × number of years) and inundation. Discrete station data were used to produce the baseline maps for the years 2025 and 2075.

The changes in shoreline location by scenario for each year are estimated as the net change caused by accelerated sea level rise, measured from the baseline for that year, and total change, measured from the 1979 USGS 7.5 minute topographic maps. The net and total change included only inundation for marsh shorelines. The net change on sandy beaches included inundation and erosion (projection of historical trends using Hands' (1981) relationship) due to the higher sea level. After sea level topped the current elevation of the dunes, the shoreline retreat was projected as a washover process, using averaged rates from existing washover islands along the South Carolina coast (as determined by Stephen et al., 1975). The total change was a summation of the historical trends and the sea level rise-induced changes. Therefore, rather than project the total disappearance of the barrier island, it was assumed that waves would build washover ridges to the spring tidal level for a uniform width which would migrate landward. The appendix tables at the end of this chapter show predicted shoreline changes for all scenarios and stations, giving a breakdown of the various components contributing to the change.

Example Analysis. As an example, the analysis for one station (52 on Figure 4-1) follows (see also appendix). The historical trend at that station for the last 40 years has been +0.3 m (1.0 ft/yr) of accretion. (It is a beach along a recurved spit on Sullivans Island.) To determine the change in the position of the shoreline for the year 2025 without accelerated sea level rise (the baseline position), the yearly depositional rate was multiplied by 45, equal to 14 m (45 ft) of accretion. Historical sea level rise rates were also projected to the year 2025 to determine the elevation of MSHW at that time, under the baseline scenario, which was a rise of 11 cm (0.4 ft). This placed MSHW for the year 2025 (baseline) at 1.0 m (3.5 ft) above present MSL. Computer-plotted maps of the present and 2025 baseline shoreline positions were overlaid and the change in position measured. For Station 52, there was a change of −6 m (−20 ft) due to inundation along the existing beach slope.

The total change in the 2025 baseline position, compared to the present, was the sum of both the historical trend and inundation, which in this case was equal to +7.6 m (+25 ft). The change in shoreline position for the 2025 low scenario can be measured from both the present shoreline (called total change) or from the projected baseline position (called net change), which is due solely to accelerated sea level rise. Net change was determined by summing the inundation component (from the comparison of contour positions for each MSHW elevation), which was −15 m (−50 ft) for Station 52, and a component for additional erosion due to the higher sea level, using Hands' (1981) model, which was −14 m (−45 ft). The total change from the present also included the change from the present due to historical trends in erosion or deposition. Thus, the total change for Station 52 under the 2025 low scenario was equal to −21 m (−70 ft), which is the sum of the projected baseline [+8 m (+25 ft)], plus changes due to inundation [−15 m (−50 ft)], plus the effect of accelerated erosion [−14 m (−45 ft)].

Shoreline changes due to inundation were measured at each station directly from the computer-generated contour maps for each sea level rise. The shoreline position was then altered where appropriate according to historical trends for baseline maps or erosion due to sea level rise on each scenario map. The shoreline between stations was interpolated using the shoreline type and adjacent stations as guides.

STORM SURGE ANALYSES

The next major impact of sea level rise considered was the alteration of storm surge levels in proportion to the sea level rise scenario. There may be minor factors that would tend to change the incremental rise in storm surge elevations, but these would be dwarfed by the present inaccuracies of inland surge modeling. The approach used was to elevate the selected storm surges (10-year and 100-year storms) by an amount equal to the sea level rise scenario. Although this technique is slightly conservative, by not accounting for displacement of the storm surge inland with sea level rise, there is no available model to estimate what the effects of sea level rise would be on the inland routing of the storm surge.

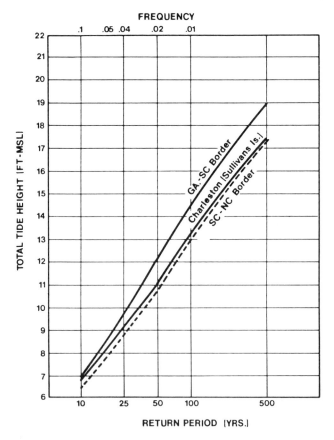

Figure 4-6. Tide frequencies at selected points on the South Carolina coast. *(After V. Myers, 1975, Storm-Tide Frequencies on the South Carolina Coast, Silver Spring, Md.: National Weather Service, Office of Hydrology.)*

Storm surge elevations for the study area were taken from Federal Emergency Management Agency (FEMA) flood maps. These maps, produced for various Charleston sites since the early 1970s, are the basis for Federal Flood Insurance rates and zoning and indicate flooding zones and corresponding surge elevations for the 100-year event (storm with a probability of 0.01). The 10-year storm elevation (with a probability of 0.1) was determined from a summary of storm tide frequencies prepared by Myers (1975) for Charleston (Figure 4-6). This figure shows that for the 10-year storm, total tidal heights would be above 1.5 m (5 ft) MSL.

GROUNDWATER ANALYSES

There have been numerous case studies of saltwater intrusion, which generally occurs from the reversal or reduction of groundwater gradients which causes denser saltwater to displace freshwater or from the destruction of natural barriers separating freshwater and saltwater. Many methods have been developed to calculate the position, simulate the motion, and predict the rate of intrusion of the freshwater/saltwater boundary (Cooper et al., 1964; Mercer et al., 1980; Pinder and Cooper, 1970). The most accurate methods involve complex convective-dispersive solute-transport equations, which require specific hydrogeological parameters and are difficult to solve. Also, for many coastal aquifers, hydrogeological parameters are not well known, not even within an order of magnitude.

A simple approach, called the Ghyben-Herzberg principle, was used as a conservative estimate of the position and change of the freshwater/saltwater boundary. The basic principle is that there is a sharp interface between freshwater and saltwater that is in hydrostatic equilibrium (i.e., no flow) due to the different densities of the two solutions. It is known that the interface is actually a broad zone of diffusion, and the saltwater is not static but flows in a cycle from the sea floor into (and creating) the zone of diffusion and back to the sea (Cooper et al., 1964). Figure 4-7 shows how this circulation pattern forms. However, the Ghyben-Herzberg principle is known to be conservative (Kohout, 1960) and can be used only as a first approximation. Only near the shoreline, where vertical flow components become pronounced, do significant errors in the position of the interface occur (Todd, 1980). Using the Ghyben-Herzberg principle, the depth to the freshwater/saltwater interface is equal to 40 times the elevation of the water table (for unconfined aquifers) or the piezometric surface (for artesian aquifers) above MSL.

There are various opinions of the effect of sea level rise on the position of the freshwater/saltwater interface in the water table aquifer using the Ghyben-Herzberg principle. On one side, the opinion is that, even though the saltwater head rises, the freshwater would also rise, and the gradients would eventually reestablish hydrodynamic equilibrium. Therefore, the whole system would shift upward in proportion to the sea level rise and landward in proportion to the shoreline retreat. The slope of the interface would control the inland excursion of the toe of the saltwater wedge beyond the new shoreline position. On the other side, the opinion is that a rise in sea level would decrease recharge (renewal of groundwater from natural resources) and increase discharge so the freshwater rise would not match sea level rise but would be some fraction of it. The increased discharge would be primarily via streams that would drain off freshwater as the water table rise intercepted the land surface. The land elevation and existing drainage patterns would determine the amount of increased discharge for a given sea level rise.

Without site-specific modeling of the groundwater flow regime, it was assumed that the freshwater/saltwater gradients in the unconfined aquifer will quickly reestablish equilibrium after sea level rise. This assumption should be valid because recharge of the aquifer is from local precipitation and is rapid through the sandy surficial sediments. The position of the saltwater/freshwater interface was calculated from the 1:40 Ghyben-Herzberg relationship. However, because the aquifer thickness averages about 13 m (40 ft), the interface will always be estimated to occur at the point where the water table is 0.3 m (1 ft) above MSL without interferences due to present water withdrawal. Using existing groundwater slopes, the position of the interface was estimated to be at approximately 60 m (200 ft) inland of the new shoreline position for each scenario. Thus, for this

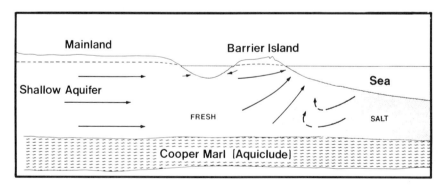

Figure 4-7. Schematic cross-section through the shallow aquifer for the eastern portion of the Charleston area showing the circulation of seawater and the general position of the zone of diffusion between freshwater and saltwater.

study, saltwater intrusion after sea level rise can be approximated by the shore erosion/inundation distance for each scenario.

For artesian aquifers, the adjustment in the freshwater/saltwater interface can be predicted using the Ghyben-Herzberg principle: that is, a 1:40 ratio for sea level rise to freshwater/saltwater interface rise (Henry, 1962). The recharge zone for artesian aquifers is generally far removed from the coast, and there would not be a significant increase in discharge. However, the time lag of saltwater intrusion is very large, as discussed in the next section.

Rates of Saltwater Intrusion

The rates of adjustment of the freshwater/saltwater zone of diffusion in groundwater in response to sea level rise will be different for water table compared to confined aquifers. Although a determination of the absolute rates is beyond the scope of this study, there are examples which demonstrate the relative rates to be expected.

There are many examples of very rapid saltwater contamination of water table aquifers due to man's activities. Large-scale construction of canals in south Florida has resulted in the penetration of saltwater into previously fresh areas—an effect somewhat analogous to sea level rise. Dense saltwater gradually replaced fresh groundwater below the canals in several years, including a drought (Parker, 1955). The saltwater zone then moved in response to gradients created by heavy pumping in the area. In New Jersey, construction of the Washington Canal in the early 1940s breached the confining layer of the shallow aquifer. By the 1980s, saltwater had traveled 8-16 km (5-10 mi) inland (Harold Miesler, 1983, USGS, personal communication). There are many other case histories that show that where shallow aquifers come in direct contact with seawater, saltwater intrusion can occur on a scale of several to tens of years. The time necessary to reach equilibrium may be much longer and is generally complicated by local changes in recharge and discharge.

The rates of adjustment in extensive artesian aquifers will be very slow, especially for the deep, stratified aquifers along the east coast. The USGS is developing a digital technique to model the movement of the saltwater/freshwater zone of diffusion during the sea level fluctuations throughout the Pleistocene epoch (Harold Miesler, 1983, USGS, personal communication). Although the model is still being developed, they estimate that the time required for stabilization of the zone of diffusion for the New Jersey sections with which they are working is on the order of 10^5 and 10^6 years.

These calculated time periods are supported by studies done by the USGS on the Atlantic continental shelf. Hathaway et al. (1979) reported that low-chlorinity water occurs beneath much of the shelf from 16 to 120 km (10-75 mi) offshore. The general pattern was described as a freshwater lens overlain by low-permeability clays, which have a sharp chlorinity gradient increasing toward seawater concentrations. They interpret the freshwater lens as a remnant of fresh groundwater that recharged the shelf sediments during the Pleistocene glacial maximum, when sea level was as much as 100 m (330 ft) lower than present. The impermeable clay has acted as a confining bed, preventing saltwater intrusion during the last flooding of the continental shelf about 8,000 years ago. Hathaway et al. (1979) proposed that the offshore freshwater lens had played an important role in preventing saltwater intrusion into mainland wellfields. The slow rates of adjustment in the freshwater/saltwater zone of diffusion is further supported by reports of remnant saline water that intruded during higher sea level stands into various coastal aquifers (Stringfield, 1966; Wilson, 1982).

RESULTS AND DISCUSSION

Smooth shoreline and flood maps for the various baseline and sea level rise scenarios for the years 2025 and 2075 were prepared from the digital terrain model and methodology already outlined. The following results offer a sampling of the changes expected under selected scenarios. A technical report by Michel et al. (1982) contains a more complete data summary.

The first set of maps prepared illustrate existing conditions, giving the location of the 1980 MSL shoreline, MSHW, and 10-year and 100-year flood zones (Figure 4-8). The maps have been combined in Figure 4-8 to illustrate the entire project area. Because of the scale at which this and subsequent maps are reproduced, it is difficult to appreciate the magnitude of many of the shoreline changes. The results indicate future shoreline change is indeed significant under all but the lowest scenarios. At the scale of these maps, a pencil width represents up to 100 m (330 ft) of change, a result that would certainly be of concern to most shorefront property owners. Despite the complexity of the maps at this scale, major trends are still apparent.

Figure 4-9 is one of the 2025 map sets that show the baseline and high-scenario position of MSHW plotted against the present MSL shoreline. Figure 4-10 similarly illustrates the predicted position of MSHW

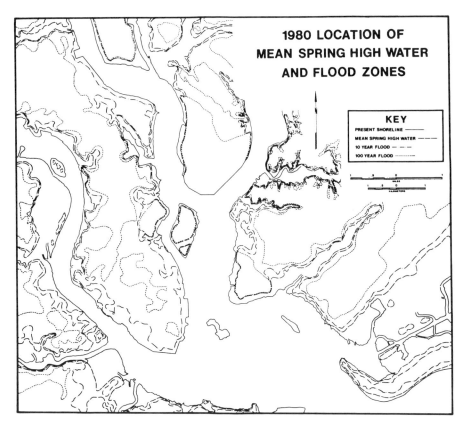

Figure 4-8. Existing (1980) locations of the MSHW, 10-year storm surge and 100-year storm surge in the study area.

for the 2075 baseline and all scenarios. These two maps illustrate the extremes in projected MSHW position for the present study. It should be obvious from a quick glance at the two maps that a very large zone of inundation would occur during the high scenario almost 100 years from now.

Figure 4-10 shows the trends in the position of MSHW for each scenario for the year 2075. On the southwestern tip of Charleston, the arrow labeled *A* represents the area of spring tidal inundation for the 2075 baseline scenario. The arrow labeled *B* represents additional areas of inundation for the low scenario; *C* represents the added area of inundation for the medium scenario; and *D* represents the additional area inundated under the high scenario.

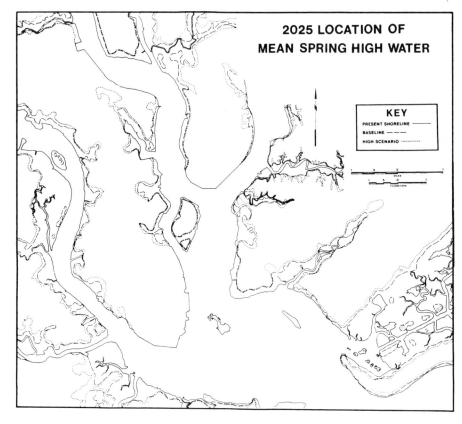

Figure 4-9. Map showing the locations of MSHW for the 2025 baseline and high scenarios. The locations for the low and medium scenarios were fairly evenly spaced between the lines shown here.

Figure 4-11 illustrates the shift in 100-year flood zones for 2075 for all scenarios, again representing the extremes for the study.

Baseline Map—Year 2025

The baseline map for 2025 (see Figure 4-9) was generated to represent the future shoreline and storm surge changes under current rates of sea level rise, which effected an 11 cm (0.4 ft) rise by 2025. When compared with existing (1980) conditions shown in Figure 4-8, there are few significant changes. An average of 30 m (100 ft) of inundation occurred along the western shore of the Ashley River, but the new MSHW was still within the astronomic tidal elevations and thus within high marsh vegetation. Along vertical seawalls and spoil dikes, the MSHW was already considered to be at the base of the structure; thus, there were no detectable changes along the man-made shorelines. The accretionary trends along the island

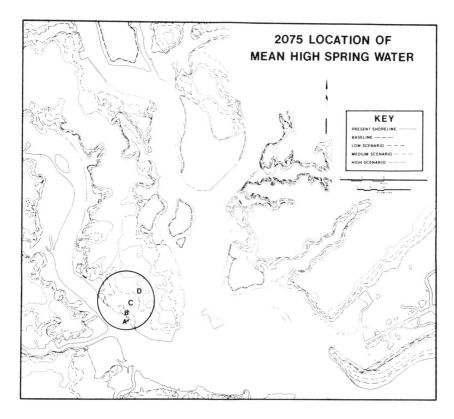

Figure 4-10. Map of the predicted locations of MSHW for the baseline, low, medium, and high scenarios for the year 2075. For diked areas, which always remained above MSHW, only the baseline and high lines are shown.

beaches dominated over the small amount of inundation. The extensive marsh between Mount Pleasant and Sullivans Island was already mostly below MSHW, except for spoil islands along the Intracoastal Waterway and areas fringing the highland. In fact, considering the accuracy of the computer-plotted contours and the ±15 m (50 ft) precision in measuring the changes between contours, there was essentially no change between present (1980) and the baseline for 2025 along interior shorelines. However, along shorelines which can be historically documented to be undergoing long-term deposition or erosion, the use of a baseline composed of a historical trend component is important. Inundation as a separate factor is not necessary because it is inherently included in the historical trend analysis.

The changes in shoreline and storm surge positions for the scenarios in 2025 were small and difficult to display at page-size scales. The shape of

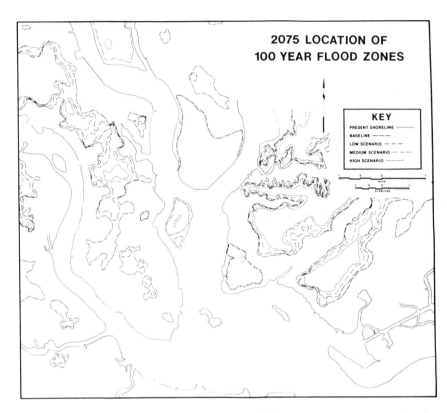

Figure 4-11. Map of the predicted locations of the 100-year storm surge for the baseline, low, medium, and high scenarios for the year 2075. For diked areas, which always remained above the 100-year surge, only the baseline and high lines were shown. Some of the diked spoil areas were affected under the medium and high scenarios.

the study area is also difficult to illustrate in sections and still retain any sense of area-wide comparisons. Thus, the results for the 2025 high scenario only are shown in Figure 4-9. The low and medium scenario results are not shown but can be visually placed between the high and baseline positions. The results are described below; the reader should refer to Figure 4-9 during the following discussion.

2025 Low Scenario

This scenario represented a total rise in sea level of 28 cm (0.9 ft) but only 17 cm (0.5 ft) above the baseline for 2025. The changes in the MSHW would be very small compared to the baseline. Inundation at the marsh stations ranged between 0 and 75 m (0-250 ft). As expected, changes in areas of narrow marshes that fringe developed highland, such as along

James Island, would not be discernible because of greater slopes (and the limitations of computer interpolation). Mount Pleasant, formed on an old barrier island itself, rises sharply above the marsh fill behind Sullivans Island; there would be little or no change in MSHW on all sides. Parts of Sullivans Island would become erosional, while the bulge in the lee of the jetties would slow its growth.

The changes in the 10- and 100-year storm surges would be small, generally less than 60 m (200 ft). A 28 cm (0.9 ft) rise obviously was not large enough to exceed any breaks in slope. The most significant change would occur on Sullivans Island, all of which is currently within the 100-year flood zone. The 10-year flood zone was predicted to dissect the island across contiguous low areas.

2025 Medium Scenario

The medium 2025 scenario of a 46 cm (1.5 ft) rise in sea level did not cause many changes in the shoreline position of consequence to developed property. At a total elevation of 1.4 m (4.6 ft) above present MSL, the new MSHW position was close to but below the 1.5 m (5 ft) contour, which is the practical lower limit for construction of permanent structures. Thus, while there would be no cases of complete structural property damage along the harbor shoreline, many structures would be placed in the zone of yearly astronomical flooding. This pattern was typical for the entire western shore of the Ashley River, which was primarily low density residential property.

Few structures would be included in the 10-year flood zone in this scenario, which ranged between 1.8 and 2.3 m (6.0-7.5 ft) above present MSL. Some new areas of residential property would be located in the 100-year flood zone, particularly between the Ashley River and Wappoo Creek (near station 11 on Figure 4-1).

The shoreline in the city of Charleston has areas that would lose up to 75 m (250 ft) due to erosion/inundation, particularly in the middle part of the peninsula. Although industrially developed, this middle section has not been landfilled to the extent which occurred to the north (U.S. Navy facilities) and south (port facilities and residential). Therefore, a narrow neck of land with smaller areas above the 10- and 100-year storm surges occurred. North Charleston, up to 10 m (33 ft) above MSL, would show even fewer shoreline changes, except along the cutbank of the Ashley River. Most of the Cooper River shoreline is composed of bulkheads and docks for the U.S. Navy Reservation and would not be affected. This area also would show regular inland shifts in the 10- and 100-year flood zones of about 75 m (250 ft).

The historical district on the Charleston peninsula had no changes along the man-made shorelines. The seawalls range in elevation between

1.5 and 2.7 m (5-9 ft) above present MSL. Thus, increasing periodicity of flooding was of more concern than inundation for this scenario. Of great importance is the projection that some of the key arteries of the city would be regularly flooded. The 10-year flood zone moved inland about 75 m (250 ft) in densely populated areas on the west side. The 100-year flood zone became scattered islands of high ground down the center of the peninsula.

Sullivans Island was the area of most serious impact. The causeway connecting the island to the mainland would be barely above spring tidal elevations. Any storm or unusual astronomical tides would regularly cut off access to and from the island. The projected position of MSHW was landward of the first row of houses in the middle section of the island. Erosion of this section would supply sediment to the western end of the island, parts of which were still accreting. Wave refraction caused by the jetties would continue to cause accretion near station 50 (Figure 4-1). Areas above the 10-year flood would be limited to a narrow strip of land down the center of the island [the only part higher than 2.3 m (7.5 ft) MSL]. Further accretion into the harbor would be limited by the deep channel and strong ebb currents, which would carry sand back out the jetties.

2025 High Scenario

There would be few additional changes in the shoreline inundation/erosion trends for this scenario, with some notable exceptions. Large movement in the MSHW position occurred on both sides of Wappoo Creek, west of the Ashley River. The scenario elevation of MSHW at 1.6 m (5.2 ft) above present MSL barely exceeded the present 1.5 m (5 ft) contour. The maps showed shoreline positions behind some existing structures and several islands of highland would be isolated in the southwestern part of the study area. Although the 10-year flood zone would get progressively larger, most of the areas above the 100-year flood west of the Ashley would now be in the flood zone.

On the peninsula, there would still be few serious shoreline problems. The newly filled and developed commercial area north of the Ashley River bridge (station 8, Figure 4-1) would be within the new intertidal zone. Otherwise, existing seawalls were high enough to prevent daily inundation. The 10-year flood would have moved 90 m (300 ft) inland of the present position. Only small areas would be above the 100-year flood along the historic district.

Currently, the town of Mount Pleasant is divided in half by a small water body (Shem Creek), which separates two highland areas. As old barrier islands, both sections are relatively high and flat. Shoreline and flood position changes would be generally small and regular, even along the convoluted areas.

Few if any structures in Mount Pleasant would be affected by shoreline movement for this scenario. The largest changes would be along the mainland facing Sullivans Island (Figure 4-9); MSHW shifted up to 225 m (750 ft) inland of its baseline position.

The causeway to Sullivans Island would be regularly flooded during spring tides, that is, every 14 days. Sullivans Island itself would have continued to narrow from both shorelines. There would no longer be any accretion on the southern end and the western tip was barely maintained by a seawall. A second row of houses would be threatened by erosion and storm waves. The 10-year flood lines would have changed little; there would still be a narrow corridor barely above the 2.5 m (8.1 ft) elevation.

Even at the highest rise for 2025, there would be no effects on the diked spoil areas throughout the harbor. Dike elevations range between 4.3 and 7.3 m (14-24 ft) above present MSL and thus would protect the spoil areas from even the 100-year storm surge.

Baseline Map—Year 2075

Projections of historical trends in shoreline position and sea level rise were used to create the 2075 baseline maps (Figure 4-10). The sea level rise of 24 cm (0.8 ft) was practically the same as the 2025 low scenario, which had a 28 cm (0.9 ft) rise. There were few areas of significant change. The new MSHW, at 1.2 m (3.9 ft), would still be below normal astronomical tides, and there would have been a gradual landward shift in marsh vegetation of a few tens of meters at the most. The only structural loss along the shoreline would have occurred in scattered locations along the seaward row of homes on Sullivans Island. Expansion of the 10- and 100-year flood zones would be highly variable but would average some 60 m (200 ft).

The changes due to sea level rise for each scenario were so large that separate maps were made for each type of coastal response. Figure 4-10 shows new shoreline positions, and Figure 4-11 shows the 100-year flood zones. Each map shows the baseline for determination of accelerated sea level rise effects.

2075 Shoreline Changes

Figure 4-10 shows the position of MSHW for the baseline, low, medium, and high scenarios for 2075. Accelerated sea level rises ranged from 0.9 to 2.3 m (2.9-7.6 ft). With an MSHW range of 0.9 m (3.1 ft), the worst-case scenario reflects an intertidal zone beginning at 3.2 m (10.7 ft) above present MSL. The low scenario exceeded in rise the highest scenario considered for 2025.

The position of MSHW for the low scenario [at 1.8 m (6 ft) above present MSL] would be inland of property along several sections of shoreline, particularly Wappoo Creek and the west shore of peninsular Charleston. Other areas of Charleston would still be protected by existing coastal structures. Sullivans Island would begin to lose a second row of houses, with 67-120 m (220-400 ft) of shoreline retreat. Under the 2075 medium scenario, the western tip of Sullivans Island would have retreated by over 600 m (2,000 ft), and the island's width would have decreased from a baseline of about 670 m (2,200 ft) to 150 m (500 ft). The island was predicted to shift to a washover mode of shoreline retreat at MSHW elevation of 2.3 m (7.5 ft). Washover islands are a flat terrace of sand which is periodically overwashed during high tides and storms. They move by landward transport of sand as opposed to alongshore transport. Dunes generally do not have time to form.

Using the 2075 high scenario, the island would have maintained its 150 m (500 ft) width and moved landward at 6 m/yr (20 ft/yr) with up to 790 m (2,600 ft) of retreat recorded.

The Mount Pleasant area would show steady shoreline inundation, with an average shift of 250 m (800 ft) in the MSHW level for the high scenario. The Charleston peninsula would have experienced the most dramatic changes in shoreline position. Under the medium and high scenarios, all existing seawalls would be overtopped, and large areas would be subsequently inundated up to 550 m (1,850 ft) for the medium scenario and 1,200 m (4,000 ft) for the high scenario. Only the central part of the peninsula would be above the intertidal zone. The entire Navy Reservation would be inundated even for the medium scenario. The spoil island dikes would still be above MSHW. Daniel Island (station 27, Figure 4-1) would become several smaller marsh islands. Highland areas west of the Ashley River would have shrunk considerably in very irregular patterns, with up to 900 m (3,000 ft) loss of land.

In summary, the areas of greatest impact would be on Sullivans Island (which became a washover island) and the Charleston peninsula (where a highly developed area underwent extensive inundation). Mount Pleasant and the spoil islands would be the least affected.

Storm Surge Elevations

With accelerated sea level rise, the 100-year flood zones would have changed dramatically. For the low scenario, only a few small patches of land would have remained above the 100-year flood west of the Ashley, on the lower peninsula, and on Daniel Island. The 100-year storm would inundate hundreds of meters of North Charleston and Mount Pleasant.

For the medium scenario, the only areas above the 100-year storm surge would be slivers of land west of the Ashley, small islands of highland in North Charleston, and two slightly smaller ridges in Mount Pleasant.

For the high scenario, the only areas above the 100-year flood elevation would be restricted to the northeastern part of North Charleston and to significantly reduced ridges in Mount Pleasant. For the first time, the diked spoil areas would show some impact—a wide part of the spoil surface would be flooded by the 100-year storm.

GROUNDWATER IMPACTS

The present position of the freshwater/saltwater interface in the water table aquifer is unknown, but it is suspected to be very close to the existing shorelines. Using the Ghyben-Herzberg relationship, the new interface was predicted to occur about 60 m (200 ft) inland of the new shoreline position. The slope of the interface should be nearly vertical because the water table aquifer is only 10-20 m (33-66 ft) deep. Therefore, the interface would eventually shift inland proportionally to the distance of shoreline inundated or eroded for each scenario. The rate of response of the interface for the water table aquifer should be close to the rate of sea level rise.

Saltwater intrusion was found to not threaten existing public water supply wells (in Mount Pleasant) until the high scenario for 2075, when the saltwater/freshwater interface was predicted to move inland 150-450 m (500-1,500 ft).

The ultimate impact of sea level rise may be negligible, considering the long-term trend for shallow coastal aquifers for the last 50 years, which has been toward a declining use and reliance on shallow groundwater. Even without accelerated sea level rise, the shallow aquifers will be overpumped, resulting in much more severe saltwater intrusion than predicted here. In 50 years, saltwater intruded up to 13 km (8 mi) in the shallow aquifer near Miami because of construction of drainage canals and heavy utilization (Kohout, 1960). On Long Island, New York, the freshwater/saltwater interface advances 3-60 m (10-200 ft) per year, depending on local pumping conditions (Todd, 1980). In the study area, Mount Pleasant pumps many of its shallow wells dry in the summer and will eventually be forced to drill more deep wells long before sea level rise becomes a factor. Additionally, shallow coastal aquifers are very prone to contamination by septic tanks, tile fields, agricultural practices, and other disposal problems. Thus, as the coastal areas become more populated, the shallow aquifers will be frequently abandoned as sources of potable water. Therefore, it is concluded that there will be no discernible effects on shallow groundwater from accelerated sea level rises in the

Charleston study area. This is not to say there is no groundwater problem, only that it has a cause not related to sea level rise. In addition, there will not be any effects on confined aquifers because the time periods necessary to reestablish equilibrium are on the order of tens of thousands of years.

ANALYSIS OF METHODOLOGY

Precision of Results

The computer-generated contour maps used in this study were made from high concentrations of digital elevation data from which contours could be plotted for specific elevations. This procedure was superior to hand interpolation between the normal 5 ft contours on the standard 7.5 minute USGS topographic maps. Even so, frequent corrections were necessary during construction of the baseline and sea level rise maps to make them conform to the USGS maps. For instance, the computer-generated maps were unable to plot accurately straight stretches of shorelines where seawalls occurred. These corrections were easily made and were not significant sources of error. The areas of greatest concern were marsh elevations, which are important for evaluation of the small sea level rises. The addition of spot elevations from large-scale maps for the marshes was critical in the generation of accurate contours between 0 and 5 ft. Even with this added detail, many manual corrections were required. To generate accurate maps at the requested detail used in this study in a routine fashion, alternative methods were necessary. The smaller sea level rises considered here were at the limit of the technique used. The digital data base needs to be even more precise than that used by the USGS to construct the base maps for accurate interpretation.

The uncertainty in the position of the predicted shorelines for the maps was at best ±15 m (±50 ft), based solely on errors due to manual transfer and line thickness. Much larger errors are possible from determination of historical trends from aerial photographs, criteria used to apply or modify the historical shoreline change rates, and interpolation of the shoreline between stations. These errors are impossible to quantify; they are a function of the data base and the judgment of the user.

Evaluation of Groundwater Analysis

The long time period for impact on confined aquifers eliminates them from consideration in this study. However, the water table aquifers are susceptible to increased saltwater intrusion. The methods used to analyze the effects of sea level rise on the water table aquifers were simple approximations of complex systems. The more precise methods, such as numerical models, require much data that are not generally available or accurately known. Even the USGS models to simulate the movement of the freshwater/saltwater interface during Pleistocene sea level fluctuations, in a region with an extensive data base, have been extremely difficult to calibrate.

The shallow aquifer in the study area was only 10-20 m (33-66 ft) thick; thus, the Ghyben-Herzberg principle predicted 60 m (200 ft) of saltwater intrusion beyond the new shoreline position for each scenario. In thicker aquifers, the Ghyben-Herzberg principle works well as a conservative estimate. The main uncertainties in its application are the degree to which the freshwater system equilibrates with the rise in saltwater head and the net effect of increased discharge. Since little is known about how these two processes affect the response of the water table, they have not been incorporated into this study. However, groundwater effects from sea level rises up to 200 cm (6.5 ft) appear to be minor compared with other processes that are causing more rapid and extreme saltwater intrusion. Studies should be made to test the impact of sea level rise on large water table aquifers that are well understood, such as the Long Island glacial aquifer, to determine if groundwater effects are an important consideration to evaluate.

General Applicability

The methods developed in this pilot study used data that are readily available for most coastal regions (i.e., various scales of topographic maps, aerial photographs, flood-hazard boundary maps) and thus are widely applicable. The methods used to predict the position of the shoreline for the baseline and scenario maps have been described in detail in this report. They are based on general principles of coastal geology and can be applied to almost any shoreline type or location. The general applicability of this method should be tested in other areas, especially to test for differences in geomorphology, tide regime, and local effects such as high subsidence rates. The coastal geomorphology and physical setting of the Chesapeake area, for example, may require a very different ordering of the dominant processes. The tidal range is smaller, and it borders a major estuary. The sediment flux will be smaller for both fine-grained, suspended sediments and littoral sediments eroding from the headlands in the bay and at the entrance capes.

REFERENCES

Brown, P. J. 1976. "Variations in South Carolina Coastal Morphology." In M. O. Hayes and T. W. Kana, eds., *Terrigenous Clastic Depositional Environments.* Guidebook to field trip sponsored by the American Association of Petroleum Geologists, pp. II-2-II-15.

Bruun, P. 1962. "Sea-Level Rise as a Cause of Shore Erosion." *Journal of the Waterways and Harbors Division* 88(WW1):117-130.

Colquhoun, D. J., T. A. Bond, and D. Chappel. 1972. "Santee Submergence: Example of Cyclic Submerged and Emerged Sequences." *Geological Society of America Memoir 133*, pp. 475-496.

Cooper, H. H., F. A. Kohout, H. R. Henry, and R. E. Glover. 1964. *Sea Water in Coastal Aquifers.* U.S. Geological Survey Water-Supply Paper 1613-C.

Finley, R. J. 1981. *Hydraulics and Dynamics of North Inlet, South Carolina, 1974-1975.* GITI report no. 10. Fort Belvoir, Va.: Coastal Engineering Research Center.

Hands, E. B. 1981. *Predicting Adjustments in Shore and Offshore Sand Profiles on the Great Lakes.* CERC technical aid 81-4. Fort Belvoir, Va.: Coastal Engineering Research Center.

Hathaway, J. C., C. W. Poag, P. C. Valentine, R. E. Miller, D. M. Schultz, F. T. Manheim, F. A. Kahout, M. E. Bothner, and D. A. Sangrey. 1979. "U.S. Geological Survey Core Drilling on the Atlantic Shelf." *Science* 206(4418):515-527.

Henry, H. R. 1962. *Transitory Movements of the Salt-Water Front in an Extensive Artesian Aquifer.* U.S. Geological Survey Professional Paper 450-B, pp. 1387-1388.

Herzberg, A. 1961. "Die Wasserversorgung Einiger Nordseebader, Munich." *Journal Gasbeleuchtung und Wasserversorgung* 44:815-819, 842-844.

Hicks, S. D. 1978. "An Average Geopotential Sea Level Series for the United States." *Journal of Geophysical Research* 83(C3):1377-1379.

Hicks, S. D., and J. E. Crosby. 1974. *Trend and Variability of Yearly Mean Sea Level, 1893-1972.* NOAA technical memorandum NOS-13. Rockville, Md.: Department of Commerce.

Hicks, S. D., H. A. Debaugh, Jr., and L. E. Hickman, Jr. 1983. *Sea Level Variations for the United States, 1855-1980.* NOAA report. Rockville, Md.

Jelesnianski, C. P. 1972. *SPLASH (Special Program to List Amplitudes of Surges from Hurricanes) 1. Landfall Storms.* Silver Spring, Md.: NOAA technical memorandum NWS TDL-46.

Jelesnianski, C. P., and J. Chen. 1984 (in press). *SLOSH (Sea, Lake, and Overland Surges from Hurricanes).* NOAA Technical Memorandum. Silver Spring, Md.: NOAA.

Kana, T. 1977. "Suspended Sediment Transport at Price Inlet, S.C." In *Proceedings of Coastal Sediments '77.* New York: American Society of Civil Engineers, pp. 366-382.

Kohout, F. A. 1960. "Cyclic Flow of Salt Water in the Biscayne Aquifer of South-Eastern Florida." *Journal of Geophysical Research* 65(7):2133-2141.

Kraft, J. C. 1971. "Sedimentary Facies Patterns and Geologic History of a Holocene Marine Transgression." *Bulletin of the Geological Society America* 82:2131-2158.

Landers, H. 1970. "Climate of South Carolina." In *Climates of the States: South Carolina, Climatography of the United States.* Asheville, N.C.: 6038, ESSA, Environmental Data Service.

Leatherman, S. P. 1977. "Overwash Hydraulics and Sediment Transport." In *Proceedings of Coastal Sediments '77.* New York: American Society of Civil Engineers, pp. 135-148.

Mercer, J. W., S. P. Larson, and C. R. Faust. 1980. "Simulation of Saltwater Interface Motion." *Ground Water* 18(4):374-385.

Michel, J., T. W. Kana, and M. O. Hayes. 1982. *Hypothetical Shoreline Changes Associated with Various Sea-Level Scenarios for the United States: Case Study, Charleston, South Carolina.* Report to ICF under contract to EPA. Columbia, S.C.: RPI.

Myers, V. 1975. *Storm-Tide Frequencies on the South Carolina Coast.* Silver Spring, Md.: National Weather Service, Office of Hydrology.

Nummedal, D., G. F. Oertel, D. K. Hubbard, and A. C. Hine. 1977. "Tidal Inlet Variability: Cape Hatteras to Cape Canaveral." In *Proceedings of Coastal Sediments '77.* New York: American Society of Civil Engineers, pp. 543-562.

Parker, G. G. 1955. "Salt-Water Contamination of the Aquifer from Tidal Canals." In *Water Resources of Southeastern Florida.* U.S. Geological Survey Water-Supply Paper 1255, pp. 682-707.

Pinder, G. F., and H. H. Cooper, Jr. 1970. "A Numerical Technique for Calculating the Transient Position of the Saltwater Front." *Water Resources Research* 6(3):875-882.

Research Planning Institute. Unpublished nearshore survey data for Isle of Palms, Kiawah Island, and Seabrook Island, 1974-83. Columbia, S.C.: RPI.

South Carolina Water Resources Commission. 1979. *Cooper River Controlled Low-Flow Study.* Columbia, S.C.: Report no. 131.

Stephen, M. F., P. J. Brown, D. M. FitzGerald, D. K. Hubbard, and M. O. Hayes. 1975. *Beach Erosion Inventory of Charleston County, S.C.: A Preliminary Report.* Columbia, S.C.: University of South Carolina, South Carolina Sea Grant technical report no. 4.

Stringfield, F. T. 1966. *Artesian Water in Tertiary Limestone in the Southeastern States.* Washington, D.C.: U.S. Geological Survey Professional Paper 517.

Todd, D. K. 1980. *Groundwater Hydrology.* New York: John Wiley & Sons.

U.S. Army Corps of Engineers. 1966. *Survey Report on Cooper River, S.C. (Shoaling in Charleston Harbor).* U.S. Army Corps of Engineers, Charleston, S.C., District.

U.S. Army Coastal Engineering Research Center. 1977. *Shore Protection Manual.* 3 vols. Washington, D.C.: Superintendent of Documents.

U.S. Department of Commerce. 1979. *State of South Carolina Coastal Zone Management Program and Final Environmental Impact Statement.* Rockville, Md.: NOAA.

U.S. Department of Commerce. 1981. *Tide Tables, North America.* Rockville, Md.: NOAA.

Ward, L. G., and D. D. Domeracki. 1979. "Hydrodynamic and Sedimentologic Processes in Tidal Channels." In *Proceedings with Abstracts, International Meeting Holocene Marine Sediments.* North Sea Basin, Texel, Holland.

Wilson, W. E. 1982. *Estimated Effects of Projected Ground-Water Withdrawals on Movement of the Saltwater Front in the Floridan Aquifer, 1976-2000, West-Central Florida.* U.S. Geological Survey Water-Supply Paper 2189.

APPENDIX

Shoreline Locations by Scenario for 2025 and 2075

Table 4-A. Shoreline Location by Scenario for the Year 2025

| Station Number | Baseline | | | Change in Shoreline Scenario (ft) | | | | | | | | | | | | | | | |
| | | | | Low | | | | Medium | | | | | High | | | | | |
	tr	in	bc	in	er	nc	tc	bc	in	er	nc	tc	bc	in	er	nc	tc
1	0	−100	−100	−250	0	−250	−350	−100	−250	0	−250	−350	−100	−350	0	−350	−450
2	−150	0	−150	0	0	0	−150	−150	0	0	0	−150	−150	−50	0	−50	−200
3	0	0	0	−100	0	−100	−100	0	−100	0	−100	−100	0	−100	0	−100	−100
4	−100	0	−100	0	0	0	−100	−100	0	0	0	−100	−100	−50	0	−50	−150
5	0	−50	−50	−50	0	−50	−100	−50	−150	0	−150	−200	−50	−150	0	−150	−200
6	0	−50	−50	−50	0	−50	−100	−50	−50	0	−50	−100	−50	−100	0	−100	−150
7	0	0	0	−200	0	−200	−200	0	−250	0	−250	−250	0	−350	0	−350	−350
8	0	0	0	0	0	0	0	0	0	0	0	0	0	−50	0	−50	−50
9	0	−100	−100	0	0	0	−100	−100	−150	0	−150	−250	−100	−950	0	−950	−1050
10	0	0	0	0	0	0	0	0	0	0	0	0	0	−150	0	−150	−150
11	0	−50	−50	−100	0	−100	−150	−50	−150	0	−150	−200	−50	−200	0	−200	−250
12	0	−50	−50	−200	0	−200	−250	−50	−400	0	−400	−450	−50	−450	0	−450	−500
13	0	−150	−150	−50	0	−50	−200	−150	−100	0	−100	−250	−150	−100	0	−100	−250
14	−14	−100	−114	−50	0	−50	−164	−114	−100	0	−100	−214	−114	−100	0	−100	−214
15	0	−100	−100	0	0	0	−100	−100	−50	0	−50	−150	−100	−50	0	−50	−150
16	—	—	—	—	—	—	—	—	—	—	—	—	—	—	—	—	—
17	0	0	0	0	0	0	0	0	0	0	0	0	0	0	0	0	0
18	0	0	0	0	0	0	0	0	0	0	0	0	0	0	0	0	0
19	0	0	0	0	0	0	0	0	0	0	0	0	0	0	0	0	0
20	0	0	0	0	0	0	0	0	0	0	0	0	0	0	0	0	0
21	0	0	0	0	0	0	0	0	0	0	0	0	0	0	0	0	0
22	0	0	0	0	0	0	0	0	−100	0	−100	−100	0	−100	0	−100	−100

(continued)

Table 4-A. (continued)

Station Number	Baseline			Low				Medium					High				
	tr	in	bc	in	er	nc	tc	bc	in	er	nc	tc	bc	in	er	nc	tc
23	0	0	0	0	0	0	0	0	0	0	0	0	0	0	0	0	0
24	0	-50	-50	-100	0	-100	-150	-50	-100	0	-100	-150	-50	-150	0	-150	-200
25	0	0	0	0	0	0	0	0	0	0	0	0	0	0	0	0	0
26	0	0	0	-100	0	-100	-100	0	-100	0	-100	-100	0	-150	0	-150	-150
27	0	0	0	0	0	0	0	0	0	0	0	0	0	0	0	0	0
28	0	-50	-50	0	0	0	-50	-50	-50	0	-50	-100	-50	-100	0	-100	-150
29	0	0	0	0	0	0	0	0	0	0	0	0	0	0	0	0	0
30	0	0	0	0	0	0	0	0	0	0	0	0	0	0	0	0	0
31	0	0	0	0	0	0	0	0	0	0	0	0	0	0	0	0	0
32	0	0	0	0	0	0	0	0	0	0	0	0	0	0	0	0	0
33	0	0	0	0	0	0	0	0	0	0	0	0	0	0	0	0	0
34	0	0	0	0	0	0	0	0	0	0	0	0	0	0	0	0	0
35	0	0	0	0	0	0	0	0	0	0	0	0	0	0	0	0	0
36	0	0	0	0	0	0	0	0	0	0	0	0	0	0	0	0	0
37	0	0	0	0	0	0	0	0	0	0	0	0	0	0	0	0	0
38	0	0	0	0	0	0	0	0	0	0	0	0	0	0	0	0	0
39	0	0	0	0	0	0	0	0	0	0	0	0	0	0	0	0	0

Change in Shoreline Scenario (ft)

(continued)

Note: Erosion is designated by negative numbers. Accretion is designated as positive numbers, without a sign. See Figure 4-1 for station locations and Table 4-2 for station descriptions.

tr = change caused by extrapolation of past shoreline trends

in = change caused by inundation due to projected sea level rise

bc = total change in baseline from 1980, equal to tr + in

er = change caused by erosion due to accelerated sea level rise

nc = net change caused by accelerated sea level rise equal to in + er

tc = total change equal to bc + in + er

145

Table 4-A. (continued)

Change in Shoreline Scenario (ft)

Station Number	Baseline			Low				Medium					High				
	tr	in	bc	in	er	nc	tc	bc	in	er	nc	tc	bc	in	er	nc	tc
40	0	-50	-50	0	0	0	-50	-50	-100	0	-100	-150	-50	-100	0	-100	-150
41	0	0	0	-150	0	-150	-150	0	-200	0	-200	-200	0	-250	0	-250	-250
42	-230	0	-230	0	0	0	-230	-230	0	0	0	-230	-230	0	0	0	-230
43	0	-50	-50	0	0	0	-50	-50	0	0	0	-50	-50	0	0	0	-50
44	0	-50	-50	-200	0	-200	-250	-50	-400	0	-400	-450	-50	-550	0	-550	-600
45	0	-50	-50	-50	0	-50	-100	-50	-50	0	-50	-100	-50	-50	0	-50	-100
46	0	0	0	-50	0	-50	-50	0	-100	0	-100	-100	0	-150	0	-150	-150
47	0	0	0	0	0	0	0	0	0	0	0	0	0	0	0	0	0
48	0	-50	-50	0	0	0	-50	-50	-250	0	-250	-300	-50	-250	0	-250	-300
49	20	-20	0	0	-45	-45	-45	0	-75	-75	-150	-150	0	-100	-105	-205	-205
50	475	-20	455	-50	-45	-95	360	455	-75	-75	-150	305	455	-400	-105	-505	-50
51	70	-20	50	-50	-45	-95	-45	50	-75	-75	-150	-100	50	-150	-105	-255	-205
52	45	-20	25	-50	-45	-95	-70	25	-75	-75	-150	-125	25	-100	-105	-205	-180
53	255	-20	235	-50	-45	-95	140	235	75	75	-150	85	235	-200	-105	-305	-70

Note: Erosion is designated by negative numbers. Accretion is designated as positive numbers, without a sign. See Figure 4-1 for station locations and Table 4-2 for station descriptions.

tr = change caused by extrapolation of past shoreline trends
in = change caused by inundation due to projected sea level rise
bc = total change in baseline from 1980, equal to tr + in
er = change caused by erosion due to accelerated sea level rise
nc = net change caused by accelerated sea level rise equal to in + er
tc = total change equal to bc + in + er

146

Table 4-B. Shoreline Location by Scenario for the Year 2075

Station Number	Baseline			Low				Medium					High				
	tr	in	bc	in	er	nc	tc	bc	in	er	nc	tc	bc	in	er	nc	tc
1	0	−250	−250	−350	0	−350	−600	−250	−450	0	−450	−700	−250	−1050	0	−1050	−1300
2	−320	−100	−450	−100	0	−100	−520	−420	−250	0	−250	−670	−420	−250	0	−250	−700
3	0	0	−200	−250	0	−250	−250	0	−700	0	−700	−700	0	−900	0	−900	−900
4	−170	−50	−220	−150	0	−150	−370	−220	−400	0	−400	−620	−220	−800	0	−800	−1020
5	0	−50	−50	−350	0	−350	−400	−50	−350	0	−350	−400	−50	−1050	0	−1050	−1100
6	0	−50	−50	−200	0	−200	−250	−50	−450	0	−450	−500	−50	−1100	0	−1100	−1150
7	0	−200	−200	−350	0	−350	−550	−200	−650	0	−650	−850	−200	−950	0	−950	−1150
8	0	0	0	−450	0	−450	−450	0	−1900	0	−1900	−1900	0	−3300	0	−3300	−3300
9	0	0	0	−1300	0	−1300	−1300	0	−1400	0	−1400	−1400	0	−1850	0	−1850	−1850
10	0	0	0	−150	0	−150	−150	0	−2300	0	−2300	−2300	0	−5450	0	−5450	−5450
11	0	−100	−100	−900	0	−900	−1000	−100	−950	0	−950	−1050	−100	−1350	0	−1350	−1450
12	0	−200	−200	−450	0	−450	−650	−200	−650	0	−650	−850	−200	−2100	0	−2100	−2300
13	0	−50	−50	−400	0	−400	−450	−50	—[a]	0	—[a]	—[a]	−50	—[a]	0	—[a]	—[a]
14	−20	−50	−70	−100	0	−100	−170	−70	—[a]	0	—[a]	—[a]	−70	—[a]	0	—[a]	—[a]
15	0	0	0	−50	0	−50	−50	0	−100	0	−100	−100	0	—[a]	0	—[a]	—[a]

(continued)

Note: Erosion is designated by negative numbers. Accretion is designated as positive numbers, without a sign.

tr = change caused by extrapolation of past shoreline trends
in = change caused by inundation a due to projected sea level rise for baseline or scenario
bc = total change in baseline from 1980
er = change caused by accelerated sea level rise
nc = net change caused by accelerated sea level rise equal to in + er
tc = total change equal to bc + in + er

[a]Shoreline completely inundated/eroded on map.
[b]Beaches now in washover mode; inundation not a factor.

147

Table 4-B. (continued)

Station Number	Baseline			Low				Change in Shoreline Scenario (ft) Medium					High				
	tr	in	bc	in	er	nc	tc	bc	in	er	nc	tc	bc	in	er	nc	tc
16	–	–	–	–	–	–	–	–	–	–	–	–	–	–	–	–	–
17	0	0	0	0	0	0	0	0	0	0	0	0	0	0	0	0	0
18	0	0	0	0	0	0	0	0	-1400	0	-1400	-1400	0	-2100	0	-2100	-2100
19	0	0	0	0	0	0	0	0	-500	0	-500	-500	0	-600	0	-600	-600
20	0	0	0	0	0	0	0	0	-1850	0	-1850	-1850	0	-2150	0	-2150	-2150
21	0	0	0	0	0	0	0	0	-1100	0	-1100	-1100	0	-3350	0	-3350	-3350
22	0	0	0	0	0	0	0	0	0	0	0	0	0	0	0	0	0
23	0	0	0	0	0	0	0	0	0	0	0	0	0	0	0	0	0
24	0	-100	-100	-300	0	-300	-400	-100	-550	0	-550	-650	-100	-1800	0	-1800	-1900
25	0	0	0	0	0	0	0	0	0	0	0	0	0	0	0	0	0
26	0	-100	-100	-150	0	-150	-250	-100	-200	0	-200	-300	-100	-500	0	-500	-600
27	0	0	0	0	0	0	0	0	0	0	0	0	0	0	0	0	0
28	0	0	0	-150	0	-150	-150	0	-650	0	-650	-650	0	-2050	0	-2050	-2050
29	0	0	0	0	0	0	0	0	-1600	0	-1600	-1600	0	-2100	0	-2100	-2100
30	0	0	0	0	0	0	0	0	-1250	0	-1250	-1250	0	-3750	0	-3750	-3750

(continued)

Table 4-B. *(continued)*

Station Number	Baseline			Low				Medium					High				
	tr	in	bc	in	er	nc	tc	bc	in	er	nc	tc	bc	in	er	nc	tc
31	0	0	0	0	0	0	0	0	0	0	0	0	0	0	0	0	0
32	0	0	0	0	0	0	0	0	0	0	0	0	0	0	0	0	0
33	0	0	0	0	0	0	0	0	−1950	0	−1950	−1950	0	−2250	0	−2250	−2250
34	0	0	0	0	0	0	0	0	−1000	0	−1000	−1000	0	−2300	0	−2300	−2300
35	0	0	0	0	0	0	0	0	0	0	0	0	0	0	0	0	0
36	0	0	0	0	0	0	0	0	−100	0	−100	−100	0	−100	0	−150	−150
37	0	0	0	0	0	0	0	0	−100	0	−100	−100	0	−150	0	−150	−150
38	0	0	0	0	0	0	0	0	0	0	0	0	0	0	0	0	0
39	0	−50	0	−50	0	−50	−50	0	−100	0	−100	−100	0	−150	0	−150	−150
40	0	−150	−50	−250	0	−250	−300	−50	−600	0	−600	−650	−50	−3000	0	−3000	−3050
41	0	0	−150	−600	0	−600	−750	−150	−1050	0	−1050	−1200	−150	−1300	0	−1300	−1450
42	−370	0	−370	0	0	0	−370	−370	0	0	0	−370	−370	0	0	−370	−370
43	0	0	0	0	0	0	0	0	0	0	0	0	0	0	0	0	0
44	0	−200	−200	−550	0	−550	−750	−200	−650	0	−650	−850	−200	−1200	0	−1200	−1400
45	0	−50	−50	−100	0	−100	−150	−50	−250	0	−250	−300	−50	−500	0	−500	−550

(continued)

Note: Erosion is designated by negative numbers. Accretion is designated as positive numbers, without a sign.

tr = change caused by extrapolation of past shoreline trends
in = change caused by inundation a due to projected sea level rise for baseline or scenario
bc = total change in baseline from 1980
er = change caused by accelerated sea level rise
nc = net change caused by accelerated sea level rise equal to in + er
tc = total change equal to bc + in + er

[a]Shoreline completely inundated/eroded on map.
[b]Beaches now in washover mode; inundation not a factor.

149

Table 4-B. *(continued)*

Station Number	Baseline			Low				Medium					High				
	tr	in	bc	in	er	nc	tc	bc	in	er	nc	tc	bc	in	er	nc	tc
												Change in Shoreline Scenario (ft)					
46	0	-50	-50	-250	0	-250	-300	-50	-400	0	-400	-450	-50	-650	0	-650	-700
47	0	0	0	0	0	0	0	0	-[b]	-2100	-2100	-2100	0	-[b]	-4250	-4250	-4250
48	0	0	0	-300	0	-300	-300	0	-[b]	-[a]	-[a]	-[a]	0	-[b]	-[a]	-[a]	-[a]
49	30	0	30	-100	-150	-250	-220	30	-[b]	-550	-550	-520	30	-[b]	-1500	-1500	-1470
50	600	-50	550	-800	-150	-950	-400	750	-[b]	-1650	-1650	-900	750	-[b]	-2100	-2100	-1350
51	120	-50	70	-200	-150	-350	-280	70	-[b]	-800	-800	-730	70	-[b]	-2000	-2000	-1930
52	75	-50	25	-150	-150	-300	-275	25	-[b]	-650	-650	-625	25	-[b]	-2000	-2000	-1975
53	425	-50	375	-450	-150	-600	-225	375	-[b]	-1000	-1000	-625	375	-[b]	-2600	-2600	-2225

Note: Erosion is designated by negative numbers. Accretion is designated as positive numbers, without a sign.

tr = change caused by extrapolation of past shoreline trends
in = change caused by inundation a due to projected sea level rise for baseline or scenario
bc = total change in baseline from 1980
er = change caused by accelerated sea level rise
nc = net change caused by accelerated sea level rise equal to in + er
tc = total change equal to bc + in + er

[a] Shoreline completely inundated/eroded on map.
[b] Beaches now in washover mode; inundation not a factor.

Chapter 5

Coastal Geomorphic Responses to Sea Level Rise: Galveston Bay, Texas

Stephen P. Leatherman

INTRODUCTION

This chapter describes the geomorphic effects of projected sea level rise on low-lying coastal landforms along southeast Galveston Bay, Texas, for a range of projected sea level rise rates (baseline, low, medium and high) at particular time periods (2025 and 2075). The objective is to determine the coastal changes in response to various assumed sea level increases by these particular dates. Two categories of physical response are addressed: shoreline changes representing landward displacement of the land/water interface and changes in storm surge levels and inland inundation as a result of the projected rates of sea level rise. Groundwater changes, resulting from saltwater intrusion into coastal aquifers, were originally considered (Leatherman et al., 1983), but sea level rise was found to result in minimal effects compared to those of cultural alterations. For instance, the groundwater supplies in the Galveston area have already been polluted or overexploited. Also, land surface subsidence, largely resulting from overpumping, has been so pronounced that legislation has recently been enacted prohibiting the further development of groundwater resources (Thompson, 1982).

A section of Galveston Island and Bay was selected for this pilot study (see Figure 5-1). This portion of the Gulf coastal plain is quite low and slopes gently seaward. Therefore, a slight rise in sea level would result in significant horizontal displacement of the shoreline and storm surge envelope. Other selection criteria included microtidal environment, major Gulf Coast estuary, highly developed population centers and industrial complexes at Texas City and north Galveston Island, availability of the National Weather Service storm surge model SLOSH (Sea, Lake and

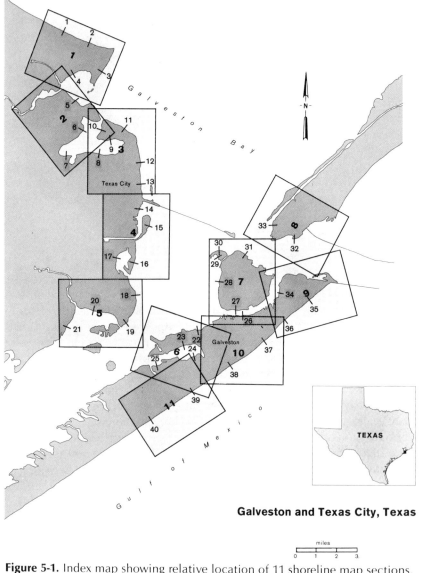

Galveston and Texas City, Texas

Figure 5-1. Index map showing relative location of 11 shoreline map sections.

Table 5-1. Accelerated Sea Level Rise Scenarios for Galveston (in cm, with ft in parentheses)

Scenario	1980	Year 2025	2075
Baseline	0	13.7 (0.5)	30.0 (1.0)
Low	0	30.7 (1.0)	92.4 (3.0)
Medium	0	48.4 (1.6)	164.5 (5.4)
High	0	66.2 (2.2)	236.9 (7.8)

Overland Surges from Hurricanes), and information on historical erosional trends and subsidence data.

Three sea level rise scenarios were developed (see Chapter 3); eight rise/year combinations were selected from the projected sea level rise curves for this analysis (see Table 5-1). The table indicates, for example, that absent any acceleration in sea level rise (the baseline scenario), by 2025 sea level will have risen by 13.7 cm (0.5 ft). In the medium sea level rise scenario, sea level will have risen by 48.4 cm (1.6 ft) in 2025. Although subsidence has been a major problem in Texas City, the estimated rate of future subsidence for this area is insignificant, with Galveston Island being essentially stable (Thompson, 1982).

As indicated in Figure 5-1, the Galveston study area was divided into several subareas. The results of this analysis are given in tables for all subareas. Graphic representations are presented for illustrative purposes in order to conserve space.

Coastal zones are inherently dynamic environments, being characterized by differing geomorphic processes and coastline configurations. To account for this wide variability in site and process, this study has combined analyses of historical trends and empirical approaches to model projected changes in the Galveston Bay area associated with the sea level rise scenarios. Shoreline changes are the major example of the former approach. Former shoreline positions portrayed on historical maps, once digitized and transformed by a sophisticated shoreline mapping program (metric mapping; Leatherman, 1983), form the basis for projecting potential shoreline excursion rates as a result of sea level rise. These extrapolated rates were assessed in light of possible impacts that recent human modification—such as levees and seawalls—may have on future trends. Empirical models have been used to derive baselines for changes in storm surge inundation. Results of the SLOSH model for the Galveston Bay area provide a base for predicting changing flood levels for each respective sea level scenario. The following sections discuss the impacts of sea level rise on shoreline retreat and storm surge levels, followed by discussions and general conclusions.

SHORELINE RESPONSE

Sea level rise has been identified as the principal forcing function in shoreline retreat along sandy coasts worldwide (Bird, 1976). As sea level rises, a number of complex and related phenomena come into play. Rising sea levels (transgression) are accompanied by a general retreat of the shoreline. This is produced by erosion or inundation. Erosion is the physical removal of beach and cliff material, while inundation is the submergence of the otherwise unaltered shoreline.

Figure 5-2 illustrates the combined effects of erosion and submergence due to sea level rise. The term D_1 represents the landward movement of the shoreline due to simple inundation of the land; the response time is instantaneous. Therefore, direct submergence of the land occurs continuously through time and is particularly evident in coastal bays where freshwater upland is slowly converted to coastal marshlands (termed upland conversion).

The second displacement term (D_2) refers to a change in the profile configuration according to Bruun (1962). The Bruun Rule provides for a profile of equilibrium in that the volume of material removed during shoreline retreat is transferred onto the adjacent shoreface/inner shelf, thus maintaining the original bottom profile and nearshore shallow water conditions (only further inland). Figure 5-3 is a more accurate depiction of this two-dimensional approach of sediment balancing between eroded and deposited quantities in an on/offshore direction. Hands (1976) found that the Bruun Rule was confirmed by actual field surveys of beach profiles during rising lake levels on Lake Michigan. The volume of sand eroded from the beach nearly matched the offshore deposition.

Beach stability in a two-dimensional sense (Bruun Rule) should theoretically be reached; W. Seelig (1982) has shown that beach equilibrium can be achieved under wave tank conditions.[1] Perhaps a constructive way of viewing the allied roles of sea level position and sea energy (coastal storms) is to consider that sea level sets the stage for profile adjustments by coastal storms. Long-term sea level rise places the beach/nearshore profile out of equilibrium, and sporadic storms accomplish the geologic work in a quantum fashion. Major storms are required to stir the bottom sands at great depths offshore and hence fully adjust the profile to the existing water level position. Therefore, the underlying assumption is that beach equilibrium will be the result of water level position in a particular wave climate setting.

Shore response lag times are tied to storm intensity and frequency, as shown by Hands (1976). The lag time in shoreline response to lake level was shown to be rather short (approximately 3 years). This rapid response

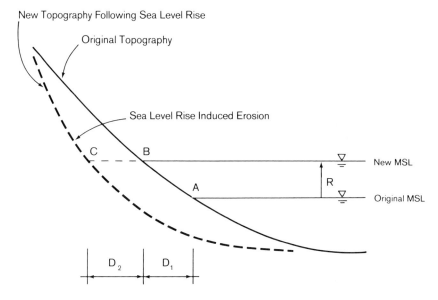

Figure 5-2. Shore adjustments with sea level rise.

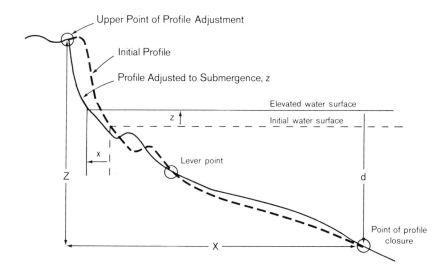

Figure 5-3. Shore adjustments to a change in water level elevation. *(After E. Hands, 1976, Predicting Adjustments in Shore and Offshore Sand Profile on the Great Lakes, CERC Technical Aid 81-4, Fort Belvoir, Va.: Coastal Engineering Research Center.)*

time is due to the fact that the Great Lakes are subject to frequent storm activity in the fall/winter before surface icing.

Along the Gulf of Mexico, sea energy is quite low except during hurricane conditions. Tropical storms are sporadic in behavior and can only be dealt with in a statistical manner (recurrence interval—a frequency/magnitude approach). Therefore, Galveston Island, for example, may be considerably out of adjustment with sea level changes over an extended period of time before being affected by a major hurricane ($\leq$ 15 years). This analysis suggests that each area will have a different lag time in shoreline response depending upon the local storm frequency, which must be treated in a probabilistic fashion.

Methods

Two different approaches can be used to model shoreline reconfiguration in response to sea level rise. The Bruun Rule describes the equilibrium profile achieved after material removed during shoreline retreat is transferred onto the adjacent shoreface/inner shelf (Bruun, 1962; Weggel, 1979; Schwartz and Fisher, 1979). The difficulty of defining the offshore limit of sediment transport limited the application of this procedure.

The second approach is less sophisticated for modeling purposes but more realistic in a geomorphic sense; it involves the empirical determination of new shorelines using trend lines. In this case, shoreline response is based on the historical trend with respect to the local sea level changes during that time period. This procedure accounts for the inherent variability in shoreline response based on differing coastal processes, sedimentary environments, and coastline exposures.

The method of projecting shoreline movement due to accelerated sea level rise is as follows. First, quantify historical shoreline movement for as long a period of record as possible (40 positions of shoreline change from 1850 to 1960 were tabulated in the study area). Second, establish a cm (ft) per year relationship for different shoreline types, wave exposures, and so on, using the historical rate of local sea level rise (Hicks et al., 1983). Last, determine a hypothesis or rule of thumb as the basis on which to project further sea level rise. In this case, shoreline movement relationships were selected, assuming that the amount of historical retreat is directly correlated with the rate of sea level rise. Therefore, a threefold rise in sea level will result in a threefold increase in the retreat rate, assuming lag effects in shoreline responses are small compared to overall extrapolation accuracy.

Tide gauge records document the relative (eustatic plus isostatic effects, such as subsidence) rate of sea level change over the period of record. The Galveston tide gauge records indicate that sea level rose about 18 cm

(0.6 ft) between 1920 and 1960 (Hicks et al., 1983). A portion of this apparent rise was probably due to subsidence: 20 cm (0.6 ft) between 1943 and 1978. However, since no future subsidence is anticipated, the long-term (100 year) rate corresponds roughly to 30 cm (1 ft) per century. If only the 1920-1940 record of sea level rise is considered, then the apparent rise would be 38 cm/100 years, of which it was estimated that 8 cm could be attributed to artificially induced subsidence. Using the longer-term data (1920-1960), a rate of 45 cm/100 years can be calculated, but at least 12 cm must be subtracted from this value because of subsidence. Therefore, an adjusted value of 30 cm/100 years has been utilized as the projected baseline sea level rise rate, since the actual amount cannot be exactly determined because of subsidence problems.

Historical shoreline data of mappable quality are available from 1850. The National Ocean Survey's (NOS, then called U.S. Coast Survey and U.S. Coast and Geodetic Survey) shoreline manuscripts were used for shoreline comparisons. The NOS T sheets were made from field surveys and are the most accurate maps of the shoreline currently available (Shalowitz, 1964). Shorelines for 1850, 1930, and 1960 were delineated and utilized in the metric mapping procedure for map compilation.

The metric mapping technique is a system that emulates the best available photogrammetric techniques by making use of the data-manipulating capabilities of modern computers. This computer package includes simple transformations of original maps to state plane coordinates, space resection of photographs to correct inherent distortions, and a method for plotting data to produce final maps. The data are input through an X-Y coordinate digitizer (Tektronics digitizer, 0.005 in. accuracy) and finally drawn by a computer-driven plotter in order to provide a spatial depiction of historical shoreline changes. This method is the most accurate technique commonly used in coastal studies (Leatherman, 1983).

The shoreline change maps for the study area (Figure 5-1) indicate significant variations in response along the shore. The geologic/environmental resources map of the Galveston area (Fisher et al., 1972) was used to determine the geomorphic/substrate units (e.g., sandy deposits, marsh, spoil, and so on). Five shoreline types were differentiated: marsh; sandy beach/bluff; enclosed water body, no significant wave action; man-made shoreline, fill; and engineering structures, particularly rubble revetments, wooden bulkheads, concrete seawalls, groins, jetties, and dikes (levees).

For sandy shorelines with no development, the maximum rate of projected retreat was applied. For marsh shorelines, the historical rate of erosion represents a conservative value for the calculation of net excursion distance because marsh drowning was not considered. For shorelines that are partially protected by engineering structures, it can be assumed that the rate of shoreline retreat will be reduced, depending upon stabilization efficiency. It should be noted that, although groins, where effective, limit longshore sediment transport by segmenting the shore-line, sand transport offshore (according to the Bruun Rule) with sea level rise can still occur. Also, groins will play little role in the abatement of shoreline recession where updrift sediment supply areas are absent or where little sand-sized material is actually being carried by the long-shore currents. Rubble toe protection is often only partially effective in limiting erosion. This stabilizing process in effect only slows down shore retreat through time as new rubble is piled along the beach in ever more landward positions.

Results

Numerical values for shoreline change were calculated from the historical maps. For example, station 2 at San Leon, which is characterized by sandy clay material, has experienced 1.1 m (3.5 ft) of erosion per year from 1850 to 1960 (see Figure 5-4). The longer-term rate is a more accurate indicator of future changes except where subsequent engineering struc-tures were certain to interfere with natural shoreline dynamics.

Future shore retreat was computed by multiplying the yearly averages by the number of years from present to the particular scenario year. Tables 5-2, 5-3, and 5-4 indicate the historical and projected shoreline changes for the various rise/rate combinations for the Galveston study area. In some cases, there are no anticipated shoreline changes due to human modifications. For example, along the Gulf Coast of Galveston Island (Figure 5-1), the shoreline is armored by a seawall erected in 1900. Therefore, no erosion is forecast for this area (see stations 34-40, Tables 5-3 and 5-4), since the beach is currently nonexistent and it was assumed that this massive engineering structure would remain intact.

In order to facilitate an economic analysis (see Chapter 7), the pro-jected shoreline changes must be related to specific area losses. There-fore, the projected recession for the eight different combinations of rise per year were manually plotted to scale on the historical (base) maps. All four projected shoreline changes for a particular year (2025 or 2075) were plotted on the base maps. For example, in the map for San Leon (Figure 5-5), the present (1960) area of landmass is indicated by screening, and the subsequent predicted shoreline positions are shown by various dash and line patterns. If the high scenario rise rate is correct, then much of the community of San Leon will be lost because of shoreline retreat.

No shore accretion is forecast for Galveston Bay or the gulfside of Galveston Island for any of the scenarios. While the jetties at Bolivar Roads Inlet have been quite effective in trapping large quantities of

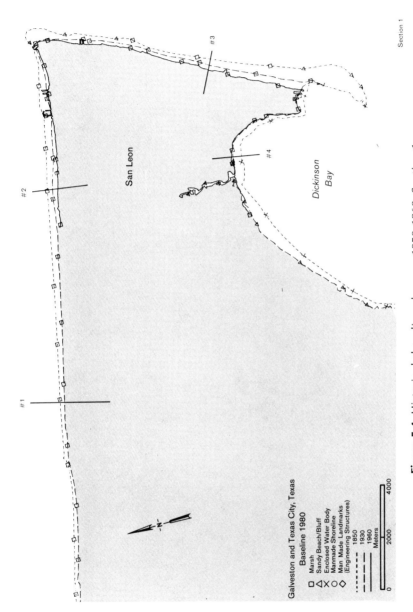

Figure 5-4. Historical shoreline changes, 1850-1960, Section 1.

159

Table 5-2. Historical Shoreline Changes, 1850–1960
(total recession in m)

Station	Material	1850–1930		1930–1960		1850–1960		
		Total	Rate	Total	Rate	Total	Rate	
1	Silt-clay	52	0.65[a]	–	–	–	–	
2	Sandy	64	0.80	61	2.03	125	1.14[a]	
3	Sandy	76	0.95	61	2.03	137	1.25[a]	
4	Silt-clay	101	1.26	15	0.51	116	1.05[a]	
5	Sandy	107	1.33	55	1.83	162	1.47[a]	
6	Silt-clay	76	0.95	52	1.73	128	1.16	0.0[a]
7	Marshy	24	0.31	–	–	–	–	0.0[a]
8	Marshy	6	0.08	24	0.81	31	0.28	0.0[a]
9	Marshy	34	0.42	3	0.10	37	0.33	0.0[a]
10	Marshy	43	0.53	3	0.10	46	0.42	0.0[a]
11	Sandy	92	1.14	55	1.83	146	1.33[b]	
12	Sandy	98	1.22	73	2.44	171	1.55[a]	
13	Sandy	52	0.65	61	2.03	113	1.03[a]	
14	Spoil	67	0.84	488[b]	–	421[b]	–	
15	Spoil	–	–	92[b]	–	–	–	
16	Marshy	40	0.50	21	0.71	61	0.55[a]	
17	Marshy	37	0.46	8[b]	–	29	0.26[a]	
18	Sandy	107	1.33	52	1.73	159	1.44[a]	
19	Sandy	79	0.99[a]	–	–	–	–	
20	Marsy	70	0.88[a]	–	–	–	–	
21	Silt-clay	67	0.84[a]	–	–	–	–	
22	Sand/spoil	168[b]	–	18[b]	–	186[b]	–	0.0[a]
23	Spoil	34[b]	–	3	0.10	31[b]	–	0.0[a]
24	Spoil	702[b]	–	0	0.00	702[b]	–	0.0[a]
25	Spoil	70[b]	0.00	45[b]	–	116[b]	–	0.0[a]
26	Spoil/bulkhead	397[b]	–	0	0.00	397[b]	–	0.0[a]
27	Bulkhead	28	0.34	0	0.00	28	0.25	0.0[a]
28	Spoil	–	–	122[b]	–	122[b]	–	0.0[a]
29	Marshy	64	0.80	15	0.51	79	0.72[a]	
30	Marshy	119	1.49	34	1.12	153	1.39[a]	
31	Sandy/spoil	–	–	76	2.54[a]	–	–	
32	Sandy/seawall	–	–	168[b]	–	–	–	0.0[a]
33	Sandy	–	–	122	4.07[a]	–	–	
34	Sandy/bulkhead	–	–	519[b]	–	–	–	0.0[a]
35	Sandy/jetty	1,373[b]	–	73[b]	–	–	–	1.0[a]
36	Sandy/seawall	183	2.29	9.2	0.31	192	1.75	0.0[a]
37	Sandy/seawall	70	0.88	0	0.00	70	0.64	0.0[a]
38	Sandy/seawall	122	1.53	0	0.00	122	1.11	0.0[a]
39	Sandy/seawall	198	2.48	89	2.95	287	2.61	0.0[a]
40	Sandy/seawall	305	3.81	24	0.81	329	2.99	0.0[a]

[a]Rate used in projection of shoreline change.
[b]Denotes accretion (in large part due to spoil deposition).

Table 5-3. Projected Shoreline Changes for 2025
(total recession in m)

Station	Descriptive Information	Baseline (13.7 cm)	Low (30.7 cm)	Medium (48.4 cm)	High (66.2 cm)
1	Groins	13	29	47	63
2	Groins	23	51	81	111
3	Groins	25	56	89	122
4	Groins	21	47	75	102
5	Fill	30	66	105	143
11	213m to dike	27	60	95	130
12	213m to dike	31	70	111	152
13	122m to dike	21	46	73	100
16	Marshy	11	25	40	54
17	Marshy	5	12	19	26
18	Sandy	29	65	103	141
19	Sandy	20	45	72	98
20	Marshy	18	40	64	87
21	Marshy	17	38	60	82
29	Marshy	15	32	51	69
30	Marshy	28	63	100	137
31	Marshy	51	114	181	248
33	Sandy	82	183	291	397

Note: Stations omitted where located along shoreline sections that have no projected changes.

Table 5-4. Projected Shoreline Changes for 2075
(total recession in m)

Station	Baseline (30.0 cm)	Low (92.4 cm)	Medium (164.5 cm)	High (236.9 cm)
1	62	188	335	481
2	108	327	583	839
3	119	360	642	925
4	100	303	540	777
5	140	424	756	1,087
11	126	381[a]	_[a]	_[a]
12	147	445[a]	_[a]	_[a]
13	98	296[a]	_[a]	_[a]
16	52	158	280	404[a]
17	25	76	135	194
18	137	415	740[a]	1,065[a]
19	94	285	508	730
20	84	255	454[a]	652[a]
21	80	242	432	622
29	68	206	367	528[a]
30	132	400[a]	713[a]	1,026[a]
31	241	730	1,301	1,873
33	387	–	–	–

[a]Indicates shoreline recession has proceeded to critical points, where further recession would result in the failure of a major structure.

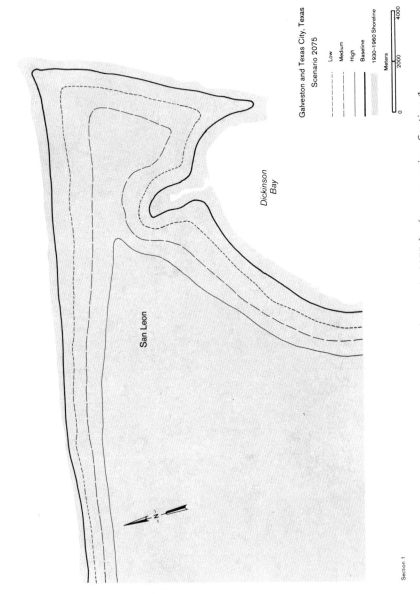

Galveston and Texas City, Texas
Scenario 2075

Low
Medium
High
Baseline
1930–1960 Shoreline

Meters

0   2000   4000

Dickinson
Bay

San Leon

Section 1

Figure 5-5. Projected shoreline changes in 2075 for four scenarios, Section 1.

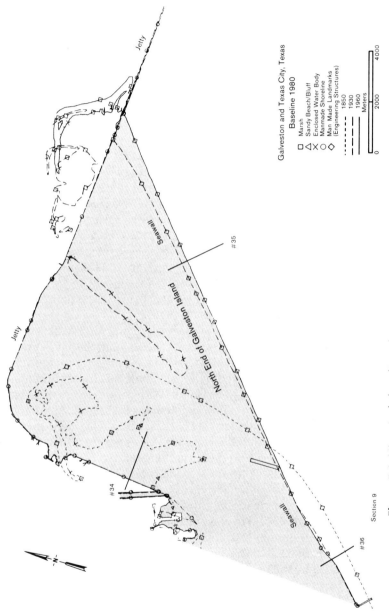

Galveston and Texas City, Texas
Baseline 1980

Marsh
Sandy Beach/Bluff
Enclosed Water Body
Manmade Shoreline
Man Made Landmarks
(Engineering Structures)

□ △ × ◇ ◇

1850
1930
1960

Meters

0   2000   4000

Jetty

Jetty

Seawall

Seawall

Seawall

North End of Galveston Island

#35

#34

#36

Section 9

N

Figure 5-6. Historical shoreline changes, 1850-1960, Section 9.

sediment derived from littoral drift on north Galveston Island (Figure 5-6), this historical trend has now been reversed, since there is little updrift sand supply. (The beach in front of the Galveston seawall is nonexistent in many locations at present.) It is assumed that continued protection to the city of Galveston will be provided by raising the height of these structures as sea level rises.

Discussion

This type of analysis could be undertaken for any coastal plain shoreline. Microtidal bays and barriers are simpler systems to model than their mesotidal counterparts with large sediment inputs from riverine sources. The easily eroded unconsolidated sediments and gently sloping, low-lying topography make the projections straightforward, except where modified by coastal engineering structures. The underlying assumption of this analysis is that shorelines will respond in similar ways in the future, as was the case in the past, since sea level rise is the driving function and all other parameters remain essentially constant. With sea level rise, Galveston Bay will tend to deepen, but this trend will be largely offset because of deposition in accordance to the Bruun Rule and continued sediment input through Bolivar Roads Inlet.

It has been assumed that all substantial engineering structures will withstand failure. With the protective beach along the Gulf Coast of Galveston nearly depleted, the seawall becomes subject to undermining. Toe protection is being provided by large rubble blocks emplaced along the seaward edge of the Galveston seawall. With continued sand depletion, however, this rubble tends to "sink" to lower levels through time so that a second line of rubble has already been emplaced yet further seaward to protect that at the seawall toe. Hence, it is necessary to build structures to protect other structures, and this situation would be further aggravated by accelerated rates of sea level rise such that a further progression of these types of very expensive activities can be forecast in the future.

This analysis has assumed that total shoreline adjustments to sea level rise would be accomplished at the particular scenario year. Clearly, there will be some lag in shoreline response to higher water levels. It is possible that this time period will be on the order of 10-15 years, corresponding to the frequency of hurricanes with a storm surge level of 1.5 m (5 ft). Since Galveston Bay depths are naturally less than 3 m (10 ft) and often less than 2 m (6 ft), a storm with such a surge would tend to overwhelm the system in comparison to the normal Bay tidal range of 0.15 m (0.5 ft) and accomplish a significant amount of the geologic work (erosion and deposition) necessary to restore profile equilibrium. Better information on storm frequency/magnitude would improve this analysis. Without site-specific data on many principal variables such as offshore profile changes, a simple extrapolation of historical trends is deemed the most reliable technique for forecasting shoreline changes.

STORM SURGE

Storm surges, the anomalously high tides produced by hurricanes and other coastal storms, are responsible for much of the damage in coastal areas as well as for extensive modification of the shoreline. The amount of damage to inland buildings and hazardous waste sites during storm conditions largely depends upon surge elevation and penetration. Sea level rise would alter storm surge levels in proportion to the amount of rise for any given scenario.

The Galveston Bay area is characterized by low-lying topography along a shallow, microtidal embayment. Sea level change will be particularly important in influencing this area, since the land is subject to flooding with even small rises. Also, storm surges superimposed on higher mean sea levels will tend to enhance shoreline erosion and bay sedimentation, as previously discussed.

Along the open coast, the amount of surge depends principally upon storm intensity and width of the continental shelf. The phase of the astronomical tide is also significant, since the coincidence of high tide and a storm surge would produce the highest storm tide (astronomical tide plus storm surge). Inland surge levels are highly variable and correspond to many variables, including basin shape and vegetation type. It is well known that surge height is amplified by funnel-shaped basins. Vegetation, particularly marshlands, attenuates the surge, since it is distributed as sheet flow over broad areas.

There are three sources of information on inland coastal storm surges: the Corps of Engineers' flood frequency curves, Federal Emergency Management Agency (FEMA) flood maps (FIRM), and the National Weather Service's SLOSH simulation computer model. Predictions of storm surge elevations are generally based on historical records of water levels occurring during previous storms. Frequency curves have been developed from a statistical analysis of tide gauge records and can be used to determine recurrence intervals for storms of particular sizes (Table 5-5). In some cases, the computation was based on frequency of central pressure indexes (U.S. Army Corps of Engineers, 1966).

Engineers and planners have established standard recurrence intervals to aid in the design of coastal engineering structures and in defining building setback lines, respectively. While various time periods have been utilized, the 100-year storm is used as the standard reference for

Table 5-5. Frequency of Storm Tide Heights for Galveston

Frequency (years)	Storm tide height above mean sea level (ft)	(m)
10	5.7	1.7
20	7.3	2.2
25	8.0	2.4
50	10.4	3.2
75	12.1	3.7
100	13.5	4.1
150	15.1	4.6

Source: From U.S. Army Corps of Engineers, Galveston District, 1966, "Texas City, Texas, Hurricane-Flood Protection," Design Memorandum no. 1, Hydrology, Galveston, Texas.

flood elevation. A 100-year flood denotes that there is a 1 percent chance of occurrence in any given year. This definition does not imply that storms of this size will be spaced precisely 100 years apart because of the probabilistic nature of frequency-magnitude relationships.

In addition to the surge frequency curves, 100-year flood maps by FEMA are available for most coastal communities. These flood insurance rate maps are essentially based on the storm surge frequency curve in combination with land elevations from U.S. Geological Survey topographic maps. The result is a map displaying the area subject to flooding during a 100-year storm. Areas so designed are further subdivided according to the potential damage, wherein the V zone corresponds to significant wave velocities encountered along the open coast. Recently, numerical storm surge models have been developed by the National Weather Service (Jelesnianski and Chen, 1983), and fortunately, Galveston Bay was one of the first areas so modeled by the National Hurricane Center using SLOSH.

Methods

The SLOSH model simulates wind speeds and storm surges based on meteorological conditions and surface characteristics (sea/bay bottom bathymetry and configuration, land elevation and morphology, and engineering structures such as jetties). This computer model numerically solves the equations of motion in order to determine surge height on a polar grid. The grid cells vary over the study area, but they are strategically placed so that the smallest cells are centered in Galveston Bay at 1.1 km (0.7 mi) spacing, with progressively larger cells being located in the Gulf of Mexico. Therefore, the polar grid permits greater accuracy (better resolution) in the more critical areas (Galveston Bay).

The SLOSH computer model yields a variety of data, but the most important in terms of storm damage functions are the composites of the storm surge envelopes that show maximum surge levels and penetrations on a grid cell basis. These surge values must have land elevations subtracted in order to yield surge penetration and height above mean sea level (see Figures 5-7 and 5-8). The three predicted flood levels corresponding to 15-20, 50- and 100- year storms have been computed for each grid square and are given in meters above mean sea level. For example, the block containing the G of Galveston (Figure 5-7) has a predicted surge of 1.5, 1.8, and 2.6 m for the three different sizes of hurricane.

The storm evacuation chart (scale of 1:62,500) by the National Ocean Survey (NOS) was utilized as the base map to overlay the transparencies by storm category. Category 3 storms (wind speeds of 111-130 miles/hour and storm center barometric pressures of 950 mb) are roughly equivalent to the 100-year storm.[2] It should be noted that these hurricane categories are for "average" conditions, and no place along the coast is really average. Thus, the actual heights of the storm tide for the 100-year storm at Galveston will exceed this table value (Table 5-5), since the surge is increased because of site-specific conditions (particularly the gentle offshore/nearshore sloping bottom). Category 1 storms (which are estimated to be 15-year storms) are the most frequently occurring event that results in significant damage. Note that the use of return intervals is somewhat subjective, since only short-term, high quality historical records are available. The 1900 hurricane, which devastated Galveston Island, was a category 3 or 4 storm; the poor quality of the available meterological data limits a precise definition of its intensity.

Over 77 separate simulations were run with the SLOSH model, based on variations in hurricane approach direction, storm speed, and differences in storm intensity along the pathway. The maximum storm surge value for each grid cell was obtained by determining the envelope for all surge overlays. The computed surges are estimated to be within 20 percent of the observed water levels (Ruch, 1981).

In interpreting the SLOSH data, the following precautions must be kept in mind: (1) integrity of the Texas City levee system and Galveston seawall is assumed; (2) no adjustments have been made for wave action that could overtop the levee and/or seawall; (3) no adjustments have been made for possible flooding by hurricane-generated rainfall behind the levee or seawall; and (4) the entire area could possibly be flooded during intense hurricanes without overtopping because of surge penetration from unprotected flanks of Texas City and bayside of Galveston Island (Ruch, 1981).

In order to determine the storm surge levels with accelerated sea level rise, the particular value for a year/rise combination was added to the

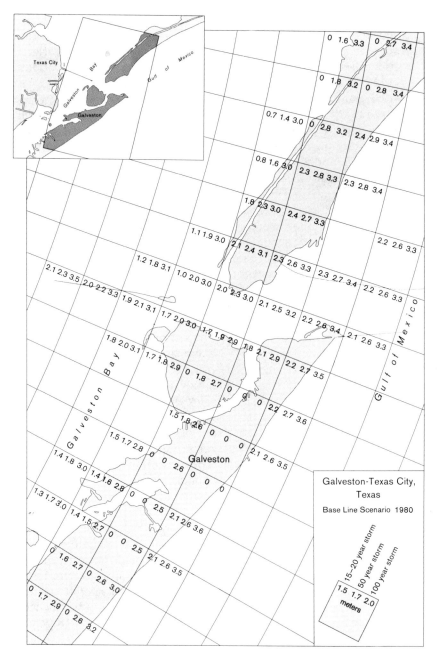

Figure 5-7. Storm surge levels for Galveston, baseline scenario, 1980.

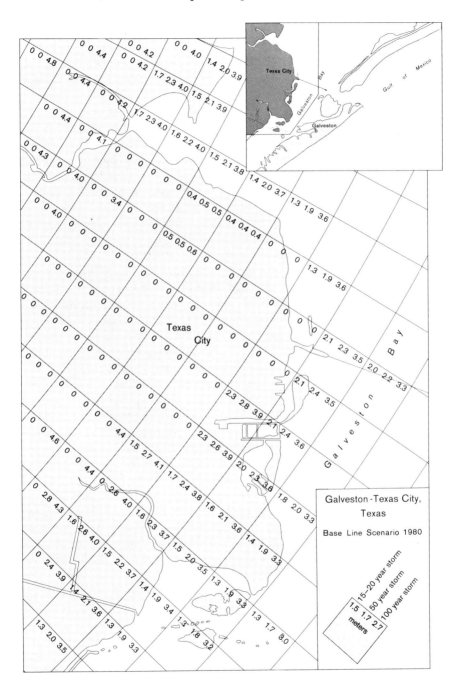

Figure 5-8. Storm surge levels for Texas City, baseline scenario, 1980.

predicted surge levels for the three storm sizes. For areas already subject to flooding, this approximation would yield a conservative value for heightened flooding, since the boundary conditions, particularly the area for storm surge ingress, were invariant. Where the SLOSH results yielded zero values for protected areas, this value was not altered unless the flood waters overtopped the threshold elevation. In this case, the algebraic sum of the maximum storm values derived by the numerical model (SLOSH) at the grid point nearest the protective structure (e.g., levee, seawall) and the height values for a particular scenario/year were used to calculate the flooding value. Since grid cell values are based on composite values from 77 computer simulation runs, it would be time and cost prohibitive to consider running this set by SLOSH for the 8 rise/year combinations (77 × 8 total permutations to calculate the maximum envelope of envelopes) to yield the actual values.

The present maximum storm surge levels predicted for 15-, 50-, and 100-year storms are shown for the city of Galveston and Texas City in Figures 5-7 and 5-8, respectively. For the SLOSH model, it was assumed that all coastal engineering structures remained intact during storm conditions and there was essentially no leakage (note zero values for the two urban enclaves). This may not be a good assumption for the Texas City levee system because this earthen structure has already been drawn close to the shore in several areas by pervasive ongoing erosion. The surge values for two small water bodies surrounded by the Texas City levee are in situ storm surge due to wind set-up. Values from Table 5-1 were used to determine the level of increased flooding due to sea level rise for areas already flooded during a particular size storm. Areas currently not flooded by a particular size storm are protected by a coastal engineering structure, such as a seawall or levee. Once the floodwaters overtop these structures, then the protected areas are subject to the same level of flooding as adjacent, nonprotected areas, provided that there is enough time to pour water over the "lip" of the structure to fill the "basin." Since the storm surge peak will only last a few hours, there may not be enough time to fill the basin, depending upon its size, the depth of overflow, and the perimeter length over which a particular depth of water is overtopping.

Galveston is subject to flooding along much of the bayside and the city is quite small in area; therefore, Galveston city would be flooded to the maximum extent possible. In the case of Texas City, flooding can often occur from both the north and south flanks. While Texas City covers a much larger area, its defenses to flooding, principally the levee system, are largely composed of earthen materials and are subject to collapse under attack by even a 50-year storm for the high scenario sea level conditions. Considering these factors, Texas City would also be flooded to the maximum extent possible once the limiting elevational criterion for flooding was exceeded.

Table 5-6. Storm Surge Flooding with Sea Level Rise for Protected Urban Areas

Location	Storm (years)	Year	Scenario	Storm Surge m	ft
Texas City	100	2075	High	6.5	(21)
Texas City	100	2075	Medium	5.8	(19)
Texas City	100	2075	Low	5.0	(16)
Texas City	100	2025	High	4.7	(15)
Texas City	50	2075	High	5.0	(16)
Texas City	50	2075	Medium	4.3	(14)
Galveston	100	2075	High	5.8	(19)
Galveston	100	2075	Medium	5.2	(17)
Galveston	100	2075	Low	3.7	(12)
Galveston	100	2025	High	3.4	(11)
Galveston	50	2075	High	5.0	(16)
Galveston	50	2075	Medium	3.5	(11)
Galveston	15	2075	High	4.4	(14)
Galveston	15	2075	Medium	3.4	(11)

Results

Table 5-6 summarizes the results of this analysis and indicates the depth of flooding for each area according to storm size and scenario/year combinations. The surge values for the entire study area have also been computed on a grid cell basis (Leatherman et al., 1983).

In the case of Texas City for a 100-year storm in 2075, flooding resulted from water rushing over low elevation areas along both the north and south flanks (see Figure 5-9). Along the north flank, the Southern Pacific Railroad tracks serve as a dike with an average maximum elevation of 3.4 m (11 ft). The floodwaters would reach this area via Dickinson Bayou in large quantities. Bayou Vista on the south flank of Texas City has a maximum continuous elevation of 4.6 m (15 ft). In all cases, the levee system along Galveston Bay, which ranges in elevation between 6.4 and 7.0 m (21-23 ft), served as a total barrier to the storm surge. Although the floodwaters, without consideration of the superimposed storm waves, would nearly reach the top of the levee, structural continuity of this coastal engineering structure was assumed.

Texas City would be subject to flooding during a 50-year storm in 2075 under the medium or high scenarios. In the case of the medium scenario, flooding would result from overtopping along the north flank only. Except for the six cases shown in Table 5-6, the water was held back by the protective devices, assuming no structural failure, and the surge values will be zero in the protective enclaves.

A similar analysis was conducted for Galveston city. Its seawall has an average elevation of 4.9 m (14 ft), effectively restraining floodwaters

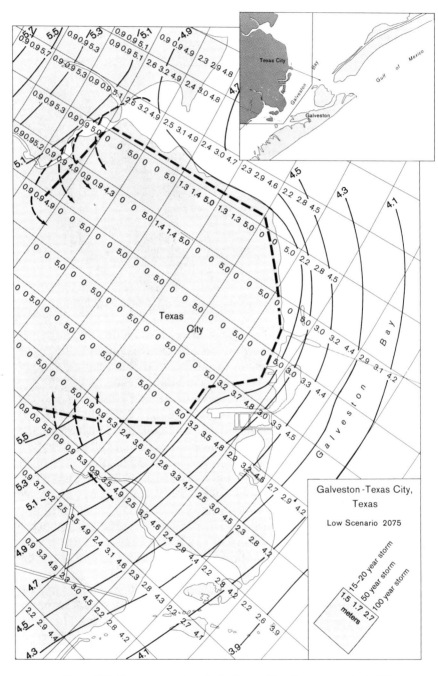

Figure 5-9. Storm surge levels for Texas City, low scenario, 2075.

below this value. Along the bay side of Galveston, the elevation varies greatly, ranging as low as 1.5 m (5 ft) to over 3 m (10 ft). While the bayside piers could be flooded under almost all conditions, it was assumed that the 3 m (10 ft) elevation was the threshold value for the flooding of downtown Galveston city. In three cases, Galveston was flooded by ocean overtopping of the seawall (2075; 100-year storm, high and medium scenarios; and 50-year storm, high scenario) as well as bayside flooding. Figure 5-10 illustrates storm surge levels for the medium scenario in 2075. In the case of oceanside flooding, it must be realized that the water is overflowing the seawall and attacking the city buildings with force because of the intensity of breaking storm waves from the deep ocean. In all other cases, flooding occurred from the bay side as the storm surge entered Galveston Bay through Bolivar Roads Inlet.

Table 5-6 is instructive in terms of planning, since it appears that the probability of damage is weighted toward the 2075 time period after sea level rise has appreciably elevated the water height. Also, the larger magnitude storms, particularly the 100-year event (class 3 hurricane), are the greatest threat in terms of flooding regardless of the particular scenario. For example, a 100-year storm can result in the flooding of both Galveston and Texas City by 2075, assuming even the low sea level rise scenario (Table 5-6).

The effect of sea level rise can also be considered in another way; the addition of even small amounts of water (sea level rise) can considerably change the storm size according to frequency (Table 5-5). The frequency of storm tide heights for Galveston Bay is based on historical records of high water levels compiled by the U.S. Army Corps of Engineers (1966). Table 5-5 represents the recurrence intervals for particular flood levels that have been statistically determined from these measurements. The general results for the entire Galveston area agree fairly well with the area-averaged values obtained from numerical modeling (Figures 5-7 and 5-8). In general, these data indicate that the 0.4 m (1.3 ft) rise in sea level associated with the medium scenario in 2025 would convert a 75-year storm into a 100-year storm (Table 5-5). Likewise, if the high scenario proved to be correct, then the flooding associated with a 100-year storm would occur at only a 10 year frequency by 2075, resulting in catastrophic damage to the study area. Also, many zones now outside the 100-year floodplain would be flooded by a 75-year storm. The potential destruction by these storm floods would significantly increase, since damage depends upon surge level.

Discussion

It is clear from the surge heights that the shape of Galveston Bay serves to amplify surge magnitude (Figures 5-7 and 5-8). In fact, open bays,

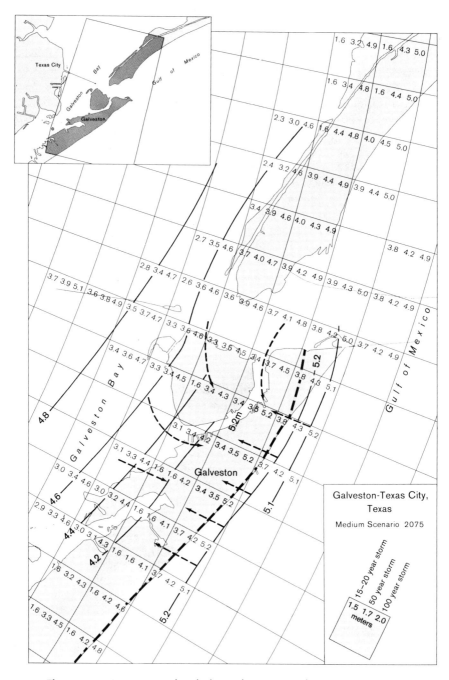

Figure 5-10. Storm surge levels for Galveston, medium scenario, 2075.

particularly funnel-shaped estuaries, tend to increase the storm surge, resulting in higher water levels during storms than those on the adjacent open coasts (Reid and Bodine, 1968).

While sea level rise would result in some modifications in bay shape and inlet configuration, these effects are too small to result in a significant change in storm surge modeling, considering the scale of the grid cell: 1.1 km (0.7 mi). Also, the storm surge analysis does not include any effects of shoreline change. Exact shoreline position is not really a factor in this analysis, considering the fact that only large-scale features are important with such a coarse grid.

There is no reason to anticipate significant alterations in major land forms due to storm surges with low levels of sea level rise by the year 2025 (except possibly for the high scenario). Galveston Bay has experienced storm surge water levels that greatly exceeded this level during this century, most notably the 1900 hurricane. With a 2.3 m (7.8 ft) rise in sea level (Table 5-1), however, modification of the shorelines of the outlying barrier islands could possibly be so drastic as to significantly alter storm surge levels. In particular, the southern end of Bolivar Peninsula could be essentially submerged or beveled (eroded) sufficiently to increase dramatically the hydraulic radius (cross-sectional area divided by the wetted perimeter) of the Bolivar Roads Inlet during storm conditions. This could greatly increase the height of the storm surge in Galveston Bay. Considerable study, however, would have to be devoted to this topic in order to determine the evolutionary characteristics and responses of low elevation landform features to such rapid rates of rise and drastic levels of the sea.

While the FEMA flood maps in conjunction with the storm frequency curve and topographic maps could have been used to yield some indication of storm surge levels, the SLOSH numerical model yields far more reliable results and allows for predictions on a grid-cell basis rather than a single averaged value for the entire basin. The National Weather Service is planning to model all the significant basins along the U.S. Atlantic and Gulf coasts, but the dates of completion for particular areas range from years to decades. This type of analysis can be utilized for any area so modeled. It should be noted that there is a fairly wide band in predicted values (accuracy within 20 percent of true value) so that ranges are more appropriate than absolute values for characterizing a storm surge.

CONCLUSIONS AND RECOMMENDATIONS

Large-scale engineering works (e.g., the Galveston seawall and Texas City levees) have been constructed to protect urban complexes on the basis of existing water level, tide, and wave energy criteria. Therefore, their resistance to failure is predicated on present conditions existing into the foreseeable future, which generally translates to a projected 100-year lifetime for most Corps of Engineers projects. It should be noted that problems have already developed along the Galveston seawall due to sand depletion, and shore erosion will soon threaten sections of the Texas City levee.

Major benefits from correctly forecasting sea level rise can be considered for two contrasting situations: undeveloped in contrast to urbanized shore. Where the land adjacent to the Gulf or Galveston Bay is undeveloped, then future facilities should be located far inland in order to avoid direct undermining with shore retreat. Building setback lines should be based on the projected shoreline positions due to accelerated sea level rise rather than a straight line extrapolation of historical trends. Also, buildings on low-lying terrain not subject to erosion will have to be placed on much higher pilings, with the height corresponding to the total rise in sea level.

Where the land is already highly urbanized, a different strategy should be advanced. Where feasible, industry should be located further inland over the longer term, instead of refurbishing old buildings. Where portside location is essential to the conduct of business, engineering structures should be built to withstand stresses generated by major storms arriving superimposed on elevated water levels. One approach is to construct heavy duty bases so that lifts can be added in the future to provide elevational adjustments in accordance with sea level rises. If basal structures are not engineered to these more rigid specifications, then the entire protective device may be ineffective during more severe conditions in the future and may not possess the strength to withstand the top-loading and reinforcement that would eventually be necessary.

In general, the principal strategy that should be employed along developed shores in the face of rising seas will involve fortressing small urban enclaves. Fortified levee and dike systems with movable locks encircling all critical facilities will prevent storm surge flooding. These fortifications will have to be well engineered to provide structural continuity.

The other alternative, which has already been suggested by the U.S. Army Corps of Engineers (1979), is to seal off the entire Galveston Bay area during storm attack. This procedure is being utilized to protect London with large movable gates that can close off the Thames River to ocean waters. After the 1938 hurricane, gates that function in a similar manner were installed along the upper part of Narragansett Bay to protect Providence, Rhode Island, from future stormwater damage. Where high relief terrain exists, such as along the glaciated New England coast, or in narrow sections that necessitate blockage, this approach will be acceptable. The suggestion has been made to emplace such a structure at the Bolivar Roads Inlet, but the cost is, at present, prohibitive and the success uncertain. In order for this type of protective device to be effective, there must be no "holes" in the dike/dam system. While the inlet gate may be made secure, the low-lying barrier islands to either side must remain intact for many miles in either direction. The higher projected rates of sea level rise may result in the partial dissolution of these barrier islands. In any case, the length of dike that would have to be maintained virtually precludes this alternative from future consideration for the Galveston area.

NOTES

1. William Seelig, 1982, Waterways Experimental Station, Vicksburg, Miss., personal communication.
2. Jarvinen, R., 1982, National Hurricane Center, personal communication.

REFERENCES

Bird, E. C. F. 1976. "Shoreline Changes during the Past Century." In *Proceedings of the 23rd International Geographical Congress*, Moscow: IGC.

Bruun, P. 1962. "Sea Level Rise as a Cause of Shore Erosion." *Journal of the Waterways and Harbors Division* 88(WW1):117-130.

Fisher, W. L., J. H. McGowen, L. F. Brown, Jr., and C. G. Groat. 1972. *Environmental Geologic Atlas of the Texas Coastal Zone: Galveston-Houston Area.* University of Texas-Austin: Bureau of Economic Geology.

Hands, E. B. 1976. *Predicting Adjustments in Shore and Offshore Sand Profiles on the Great Lakes.* CERC Technical Aid 81-4. Fort Belvoir, Va.: Coastal Engineering Research Center.

Hicks, S. D., H. A. Debaugh, Jr., and L. E. Hickman, Jr. 1983. *Sea Level Variations for the United States, 1855-1980.* NOAA report. Rockville, Md.: NOAA.

Jelesnianski, C. P., and J. Chen. 1984 (in press). *SLOSH (Sea, Lake, and Overland Surges from Hurricanes).* NOAA technical memorandum. Silver Spring, Md.: NOAA.

Leatherman, S. P. 1983. "Shoreline Mapping: A Comparison of Techniques." *Shore and Beach* 51:28-33.

Leatherman, S. P., M. S. Kearney, and B. Clow. 1983. *Assessment of Coastal Responses to Projected Sea Level Rise: Galveston Island and Bay, Texas.* URF report TR-8301; report to ICF under contract to EPA. College Park: University of Maryland.

Reid, R. O., and B. R. Bodine. 1968. "Numerical Model for Storm Surges in Galveston Bay." *Journal of the Waterways and Harbors Division* 94(WW1):33-57.

Ruch, C. 1981. *Hurricane Relocation Planning for Brazoria, Galveston, Harris, Fort Bend and Chambers Counties.* Texas A&M University Sea Grant report 81-604. College Station, Tex.: Texas A&M University.

Schwartz, M. L., and J. J. Fisher, eds. 1979. *Proceedings of the Per Bruun Symposium.* Newport, R.I.: IGU Commission on the Coastal Environment.

Shalowitz, A. L. 1964. *Shore and Sea Boundaries.* Vol. 2. Department of Commerce Publication 10-1. Washington, D.C.: Government Printing Office.

Thompson, R. E. 1982. *Subsidence '82: Harris-Galveston Coastal Subsidence District.* Houston, Tex.: Estey Houston and Associates.

U.S. Army Corps of Engineers, Galveston District. 1966. "Texas City, Texas, Hurricane-Flood Protection." Design Memorandum 1, Hydrology. Galveston, Texas: U.S. Army Corps of Engineers.

U.S. Army Corps of Engineers, Galveston District. 1979. *Texas Coast Hurricane Study: Galveston Bay Study Segment.* Galveston, Tex.: U.S. Army Corps of Engineers.

Weggel, R. 1979. *A Method for Estimating Long-Term Erosion Rates from a Long-Term Rise in Water Level.* CERC Technical Aid 79-2. Fort Belvoir, Va.: Coastal Engineering Research Center.

Chapter 6

Control of Erosion, Inundation, and Salinity Intrusion Caused by Sea Level Rise

Robert M. Sorensen, Richard N. Weisman,
and Gerard P. Lennon

INTRODUCTION

The most important direct physical effects of a significant rise in mean sea level are: coastal erosion, shoreline inundation owing to higher normal tide levels plus increased temporary surge levels during storms, and saltwater intrusion primarily into estuaries and groundwater aquifers (see Ippen, 1966; Komar, 1976; Sorensen, 1978; and Todd, 1980, for basic discussions of these phenomena). With a few exceptions, a significant sea level rise will increase the normally adverse effects of these phenomena.

In many coastal areas, economic considerations will not justify a response to these sea level rise effects. Where a response is justified, it may be political (zoning to prevent growth in areas of potential inundation and erosion), structural (building of coastal dikes to control inundation or saltwater intrusion barriers for aquifers) or, most likely, a combined political/structural response.

This chapter describes structural methods for controlling erosion, inundation, and salinity intrusion caused by sea level rise, including typical costs and the expected general effectiveness of these methods (in light of the anticipated sea level rise scenarios).[1] Both "hard" and "soft" structural responses are presented. The term hard structures refers to

179

structures such as seawalls and levees. Soft structural responses include artificial beach nourishment to counter erosion and flooding or injection of water into a well along the coast to develop a saltwater intrusion barrier in an aquifer. Both the cost and the effectiveness of any structural control method are extremely site dependent and quite variable from site to site.

The next section of this chapter covers methods for the control of erosion and inundation, while the third discusses control of salinity intrusion. Inundation is a major cause of, and is difficult to separate from, shore erosion where erosion is active; thus the two are presented together. Each section discusses, as necessary, the processes involved in coastal erosion, inundation, and salinity intrusion; the basic approaches used to control these phenomena; and details of the specific control methods including their costs and effectiveness. The final section of the chapter summarizes the key points and suggests how these control methods might be applied at a given site.

CONTROL OF EROSION, INUNDATION, AND STORM SURGE

This section discusses the effects of sea level rise on erosion and inundation, and then on storm surge, in terms of both their processes and control. Specific methods for erosion, inundation, and storm surge control are then presented.

Sea Level Rise Effect on Erosion and Inundation

Processes. The processes involved in, and the resulting extent of, shore erosion depend largely on the type of shore being eroded. The discussion of erosion processes is thus presented according to the more common types of shoreline found in the United States. Komar (1976) and Sorensen (1978) provide a general discussion of shore erosion processes.

Beaches. Except near tidal entrances, sand transport on beaches is controlled primarily by wind wave action and secondarily by wind-generated currents. Wave action can move sand in both the onshore-offshore direction and in the alongshore direction.

During storms, waves break higher up the beach profile and cut the beach face back, depositing the sand offshore. When smaller waves follow a storm, sand is moved back onto the beach face from offshore. Thus, a cyclical change in beach profile occurs due to onshore-offshore sand movement. Net recession of the shoreline is possible if sand is carried too far offshore by storms or if a sequence of above-average stormy seasons moves more sand offshore than can be returned by the milder waves during the interlude seasons.

Alongshore transport of sand occurs when waves approach the shoreline at an angle. Sand is moved along the coast in the direction of the alongshore component of wave energy. If insufficient sand is available to satisfy the transport capacity of the waves, sand will be taken from the beach to satisfy that transport capacity. Groins, jetties, and other works of man that trap sand on the beach can cause erosion in the downcoast direction from the structure.

Thus, the erosion of beaches during an essentially static sea level is caused primarily by waves carrying sand offshore during storms and by the alongshore transport of sand not being satisfied by available sand. The latter typically dominates. Winds generating currents that move sand alongshore and offshore contribute to the erosion.

With a significant rise in sea level there will be an acceleration of beach erosion in areas already eroding and possibly a start of erosion in areas not previously subject to erosion. There are several reasons for this.

1. The main reason for increased erosion is simply that the higher water level allows wave and current erosion processes to act farther up on the beach profile and cause a readjustment of that profile, which results in a net erosion of the beach and deposition on the nearshore bottom.

2. Beach profiles are concave, increasing in steepness nearer to shore. At higher sea levels, waves can get closer to shore before breaking and cause increased erosion.

3. Deeper water also decreases wave refraction and thus increases the capacity for alongshore transport.

4. Higher sea level could change the source of sediments, for example, by decreasing river transport to the sea as the mouth is flooded. However, higher sea level can also act to diminish erosion by making more material available to alongshore transport by allowing wave attack on previously untouched erodable cliffs.

In summary, the general effect of shoreline rise on a beach profile is to move the profile shoreward.

Cliffs.   Cliffed coasts often, but not always, have a thin protection beach, which may be temporarily removed during a storm, allowing wave attack at the base of the cliff and undermining of the cliff face, in turn causing a recession of the cliff. Cliffs vary in composition from extremely resistant rock that, under constant exposure to waves, shows little change in a century to loose materials that can be cut back tens of feet when attacked during a single storm. A rapidly rising sea level greatly increases both the exposure of the base of cliffs and the resulting erosion rate of erodable cliffs.

Estuaries.   Estuary shorelines are typically exposed to much milder wave action and consist of fine materials and very flat shore profiles. Rising sea levels will flood the shoreline causing greater land loss owing to inundation than that owing to erosion.

Reefed Coasts.   Many tropical coasts, for example, Oahu, have thin fragile beaches protected by offshore reefs that cause wave breaking and milder wave action on the beach. A rapidly rising sea level would increase the water depth over the reef and wave action on the beach or cliff inside the reef, resulting in increased erosion.

A rise in sea level will cause coastal inundation, an effect that is difficult to separate from the effect of shore erosion where erosion is occurring. At the water line, typical beach profiles have a slope that can vary from 1:5 to 1:100, so a 1 ft sea level rise can move the shoreline from 5 to 100 ft landward in addition to any landward movement of the profile owing to erosion.

Inundation by sea level rise will also require the raising and/or waterproofing of structures already inundated by virtue of having been built in the water. Examples include jetties built at the entrance to a navigation channel; bulkheads, docks and launching ramps in coastal marinas; and causeways across coastal embayments. Storm drainage, sewerage, and other liquid discharges to the sea by pumped or gravity flow through pipelines will have to be assisted in many areas by the installation of additional pumping capacity.

Deeper coastal waters will increase the tidal range along the coast, which will compound the above-mentioned effects of inundation and possibly change tide-induced flow and sedimentation patterns in coastal waters. Also, deeper coastal waters will permit the penetration of higher waves into coastal waters so that structures needing to be raised may also have to be strengthened to withstand increased levels of wave attack.

Control.   To prevent shoreline erosion one must keep waves, particularly storm waves, from attacking the shore by intercepting them seaward of the surf zone or by armoring that portion of the shore profile where erosion occurs. Where longshore transport is significant, erosion can be prevented by reducing the ability of waves to transport sand or by increasing the supply of sand available for transport by the waves so that sand is not removed from the beach to satisfy the sand transport capacity. Most erosion control methods act in more than one way.

Economics will justify erosion control only at selected locations, such as densely populated areas, defense installations, or sites of historic significance, many of which will already have some works to control erosion. Where shore erosion control works already exist, a response to sea level rise will often require building up the cross-sectional size and/or the stability and durability of the existing works.

Of the political responses to erosion, the eroded position could be continuously rebuilt or abandoned. New efforts to respond to erosion might include controls on further site development by appropriate agencies and abandonment of existing development. These approaches also generally apply to inundation.

In coastal regions, inundation caused by a rise in sea level and the increased tide range can be prevented by constructing a water-tight continuous structure such as the dike systems in the Netherlands; or if the inundated area is not too large, fill can be placed and held by a retaining structure. For areas of high wave attack, the dike or retaining structure will also have to be an erosion control structure. With a dike system, interior drainage canals and pumps to remove water that seeps into the areas below sea level will likely be needed.

Increased wave attack owing to higher water levels or wave attack at higher elevations because of the raised sea level will require many coastal structures to be stabilized by the addition of more or larger armor material and the armoring of areas not previously exposed to wave attack.

Higher sea levels will diminish the functionality of certain structures. For example, breakwaters that become more easily overtopped by waves would have to be raised to maintain their effectiveness; many marina and harbor appurtenances, such as fender systems and docks and walkways, would have to be raised.

Sea Level Rise Effect on Storm Surge

Processes.   A storm with high sustained winds can cause a storm surge along several types of coastline. Depending on the configuration of the coast, different mitigation techniques are used. In the discussion that follows, three types of coastline or coastal feature are discussed: long narrowing bays or estuaries, open coastlines, and wide bays or sounds.

Each class of coastal feature is subject to different types of damage when storm surge occurs. Hence, mitigation measures must be used that are best suited to the damage potential. Some solutions are common to all areas.

Long Narrowing Bays and Estuaries.   Storm surge causes much damage in funnel shaped bays, such as Narragansett Bay, Rhode Island. Water elevations can be dramatically higher at the head of the bay than those along the open coast. The hurricane of 1938 caused a surge almost 5 ft higher in Providence than at the mouth of the bay.

This phenomenon occurs in estuaries that narrow in the direction of the strongest winds. Thus, on the east coast of the United States, where damaging hurricanes move generally northward, estuaries that narrow northward such as Charleston harbor are susceptible to large surges. If the storm center passes to the west of the estuary, the counterclockwise blowing winds of the hurricane can move directly up the estuary. Typically, an urban area is situated at the head of the bay.

The high water in the bay caused by storm surge produces a backwater effect that in turn causes flooding in the tidal rivers that drain into the bay or estuary. This backwater effect can cause urban stormwater drainage systems to malfunction. Because of the heavy rain associated with most hurricanes, the backwater problem can cause extensive flood damage.

Open Coasts. An open coast, such as the barrier islands that run along much of the East and Gulf coasts of the United States or the unobstructed shoreline of the West Coast, is subject to both flooding and severe wave action during a storm surge. Although bluffs or cliffs would not flood, a wave attack at these high elevations could accelerate their erosion.

Bays, Sounds, and Harbors. In a semi-enclosed basin, such as the sounds between barrier islands and the mainland, storm surge at the open coast forces water through tidal inlets; the surge can also overtop the barrier islands. However, the limited distance over which the waves could build up over a small body of water limits the wave size. Hence, the structures can be designed for smaller waves.

Some bays and sounds may be subjected to a surge during an ocean storm if the barrier island is breached during the storm. This surge happens if the beach dune system erodes away. The storm surge moves as a wave across the sound to the mainland. A healthy wetlands or marsh system may help attenuate this storm surge.

Control. *Long Narrowing Bays and Estuaries.* Two structural mitigation methods are used to prevent flooding at the head of bays and estuaries. Usually the damage potential of waves is not as great as along the open coast. The two control techniques are a dam or tidal barrier and levees and/or floodwalls around flood-prone areas.

A tidal barrier prevents the storm surge from moving up the estuary into an urban area and also prevents the backwater effect in streams that run into the estuary. The barrier usually contains gates that are opened to allow navigation during normal weather and closed when a storm approaches. Levees and floodwalls are built in association with a tidal barrier to prevent flanking by the floodwater.

In an estuary where no tidal barrier is built, levees can be constructed to protect a flood-prone area. In either case, pumping facilities are essential to drain runoff from upland areas or water trapped inside a ring levee.

Providence, Rhode Island, and New Bedford, Massachusetts, have tidal barriers and associated levees, floodwalls, and pumping facilities to mitigate storm surge flooding (see Childs, 1965, and U.S. Army Corps of Engineers, New England District). The Thames River, east of London, also has a $1 billion tidal barrier to prevent storm surge flooding of London.

Nonstructural alternatives for storm surge mitigation in bays and estuaries include zoning, floodproofing, and abandonment of high risk areas.

Open Coasts.   Where flooding must be mitigated and wave action is severe, protective structures must be designed to withstand the forces associated with large storm waves. Also, the height of the structure must be built to an elevation that prevents frequent overtopping by the runup of breaking waves.

Open coastlines can be categorized as natural or urbanized. A natural reach of coastline typically has a beach, a line of dunes, and perhaps some roads or structures behind the dunes. Mitigation of storm surge along a natural coastline can employ beach nourishment and dune building or dune stabilization techniques.

In highly urbanized areas where the dunes and beaches no longer exist or do not offer adequate protection, seawalls and revetments must be used. These structures are designed to reflect or dissipate wave energy and are high enough to protect inland areas from flooding.

Seawalls and revetments are found at various locations along the East and Gulf coasts. The Galveston seawall, for example, was built in response to a hurricane surge that occurred at the turn of the century and killed 6,000 people.

Bays, Sounds, and Harbors.   Mainland areas behind barrier islands or land areas surrounding harbors and bays are protected from flooding by levees and floodwalls. Examples of such areas are the Galveston Bay, Matagorda, and Corpus Christi areas of Texas (see U.S. Army Corps of Engineers, Galveston District, 1979) and East Coast bays such as the Chesapeake, Delaware, and Raritan bays.

A rise in sea level in these areas will allow storm surge to flood land areas previously immune to coastal flooding or to affect some land areas more frequently. Existing structures will have to be strengthened and made more effective against a higher static water level and waves. Areas that are not now protected against storm surge flooding and wave attack will be forced to decide between abandonment and initiation of engineering works.

Specific Erosion, Inundation, and Storm Surge Control Methods

Although a rising sea would require some change in application, the methods for controlling the effects of coastal erosion, inundation, and storm surge from sea level rise would be the same as the methods employed today. These methods can be broadly classified as hard, soft, and miscellaneous. The letters in parentheses below indicate whether the method is employed primarily against erosion (E), inundation (I), and/or storm surge (S).

Hard structures include: offshore breakwaters (E), perched beach (E/I), groins (E), revetments (E/I/S), dikes (E/I/S), seawalls (E/I/S), bulkheads (E/I/S), and dams (I/S).

Soft structures include: artificial beach nourishment (E/I/S), dune building (E/I/S), and marsh building (E/I/S).

Miscellaneous inundation responses include: elevation of structures (I/S), strengthening of structures (I/S), and expanding water collection and pumping systems (I/S).

As erosion, inundation, and storm damage occur, economic considerations would dictate that the most lightly inhabited coastal areas be abandoned. Control of erosion, inundation, and storm damage would not be attempted in uninhabited and lightly inhabited coastal areas unless a site has value for some other reason, for example, an historic site or a potential defense site.

When it is apparent that a significant sea level rise is occurring, coastal political authorities will have to limit and direct development to locations that can most economically and effectively be defended. This would have to be accomplished through zoning, penalties to construction in undesirable areas (such as higher insurance rates), and perhaps even condemnation of existing development that cannot be protected.

Raising and strengthening existing structures or expanding flow discharge systems involves an enormous variety of efforts, including determining their costs, application, and effectiveness. A discussion of these methods is beyond the scope of this chapter.[2]

Offshore Breakwaters. One or more breakwaters, with intervening gaps, have been built parallel or nearly parallel to shore in water depths of a few to 20 or 30 ft to stabilize a shoreline. They function by intercepting a large portion of the incident wave energy and thereby decrease the offshore and alongshore transport capacity of waves.

Where significant alongshore transport occurs, offshore breakwaters will trap a portion of that transport to augment the original beach. If the original beach is inadequate and the potential for trapping a significant volume of sand from alongshore transport does not exist, the area in the lee of the breakwater(s) can be filled with sand. Figure 6-1 shows a section of nourished beach at Lakeview Park, Ohio, protected by three offshore detached breakwaters and a groin.

Offshore breakwaters are usually constructed when a new beach is to be developed or an existing beach is to be stabilized. A shoreline without a beach of any consequence is more likely to be stabilized by construction of a structure at the land-water interface such as seawalls, revetments, or bulkheads rather than an offshore breakwater. A tradeoff can be made between the size, length, and crest elevation of breakwaters and the resulting level of transmitted wave energy versus resulting beach shape, erosion, and consequent need for periodic renourishment. Most offshore breakwaters are built with a low crest elevation to minimize cost, maximize water circulation in their lee, and minimize beach planform irregularity.

Offshore breakwaters are quite effective in stabilizing shorelines, but, particularly on exposed coasts having higher waves, their capital cost can be quite high. The structural aspects of their design are reasonably well understood theoretically, but their functional layout, length, gap width, distance offshore, and crest elevation are generally based on empirical evidence.

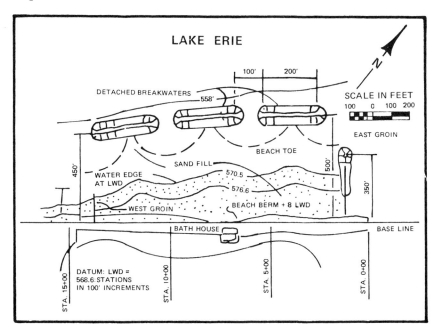

Figure 6-1. Three offshore (detached) breakwaters, a groin, and beach nourishment at Lake View Park, Ohio.

Offshore breakwaters, like other breakwaters, are typically stone rubble mounds, but they have been built with steel or concrete sheet piling, sand filled bags and rubber tubes, and wooden cribs filled with stone. To allow some wave transmission through the structure, some have been constructed from large armor stone only.

Table 6-1 lists the cost per foot of breakwater length and cost per foot of beach protected for recent offshore breakwaters. The first three sites are on the Great Lakes and the fourth is in more protected waters in Delaware Bay. No recent cost data were obtained for offshore breakwaters on the open ocean. Economic considerations diminish their use for beach protection in high wave environments. An estimate of the cost per foot of structure in the open ocean can be obtained by looking at cost figures for jetties with similar cross sections. The proposed 1,280 m (4,200 ft) long rubble mound jetty at Barnegat Inlet, New Jersey, has an estimated cost of $25,800,000 (see U.S. Army Corps of Engineers, Philadelphia District, 1981) or an average cost of $20,139/m ($6,140/ft) in 1981. Typical dimensions for this jetty are a 7 m (23 ft) height, 24 m (80 ft) base width, and 6m (18 ft) crown width. A similar offshore breakwater protecting a beach might have a cost of $9,840-$13,120/m ($3,000-$4,000/ft) of beach. Thus, the cost of offshore breakwaters in 1980 dollars could vary from $656-$9,840/m ($200-$3,000/ft) of beach depending on the level of wave attack, nearshore beach slope, and level of protection desired.

As sea level rises, the crest elevation of an existing offshore breakwater would have to be raised to maintain the same level of shore protection. Higher waves would approach the gaps owing to the deeper water, so gap widths may have to be decreased. Also, higher sea levels will increase the distance between the breakwaters and the shoreline and allow more wave energy to reach the protected area owing to wave diffraction at each end of the line of offshore breakwaters. Thus, the two end breakwaters may have to be extended. Beach fill and/or shoreline structures might be constructed landward of the breakwaters to keep the shore from being inundated. Because rubble structure costs vary geometrically with crown elevation, a 10 ft rise would generally cost much more than five times the cost of controlling a 2 ft rise.

Perched Beach. Related to offshore breakwaters but functioning in a different way is the concept of a perched beach. A continuous well-submerged structure is built offshore and parallel to shore, and a beach is built between the structure and shore by artificial nourishment. The structure retains the toe of the beach and perhaps diminishes incident wave energy somewhat by causing larger waves to break. Being submerged, the structure is not exposed to large wave forces.

A perched beach was proposed for the coast of California at Santa

Table 6-1. Offshore Breakwater Cost Data

Site	Cross-section dimensions	Cost/ft		Remarks
		Structure	*Beach*	
Presque Isle Erie, Pennsylvania	15' vertical 65' base 13.5' crown width rubble mound	$1,728	$585	Project also calls for beach fill; groin field exits at site
Lakeshore Park Ashtabula, Ohio	10' vertical 45' base 9.5' crown width rubble mound	622	330	Breakwater is all armor stone; beach fill provided
Lake View Park Lorain, Ohio	(see Figure 6-1)	890	700	Beach fill and terminal groin
Kitts Hummock Delaware	69'' Ø Longaard tube with sand fill	188	91	Structure crests at mean sea level; structures part of demonstration
	Six 4' × 20'' × 12' sandbags	216	105	project and probably somewhat
	5' vertical 20' base 5' crown width rubble mound	228	111	underdesigned

Sources: For Presque Isle: U.S. Army Corps of Engineers, Buffalo District, 1980, "Construction Cost Estimate: Beach Erosion Control Project, Presque Isle, Erie, Pennsylvania."

For Lakeshore Park: U.S. Army Corps of Engineers, Buffalo District, 1982, "Construction Contract Cost Bidding Schedule: Beach Erosion Control and Shoreline Projection Project, Lakeshore Park, Astabula, Ohio."

For Lake View Park: U.S. Army Corps of Engineers, Buffalo District, 1977, "Construction Contract Cost Bidding Schedule: Beach Erosion Control Project, Lake View Park, Lorain, Ohio."

For Kitts Hummock: U.S. Army Corps of Engineers, Philadelphia District, 1978b, "Kitts Hummock, Delaware, Pre-construction report, Shoreline Erosion Control Demonstration Program."

Monica (see Dunham, 1968) but not built. As part of a shore erosion control demonstration project in sheltered waters, Longard Tube, sand bag, and wood sheet pile structures were built at Slaughter Beach, Delaware (see U.S. Army Corps of Engineers, Philadelphia District, 1978a). Structure crests were about 0.6 m (2 ft) above the bottom in water averaging about 1.2 m (4 ft) deep and sand fill was placed behind the structures. Structure costs were $279-$515/m ($85-$157/ft) in 1978. Structure costs for a perched beach might be roughly one-fourth to one-half those of an offshore breakwater on a per foot of structure basis or one-half to three-quarters on a per foot of beach basis.

As sea level rises, the crest elevation of an existing perched beach structure would be raised, and more sand would be placed behind it. An offshore breakwater system being submerged by a rising sea level could be converted to a perched beach system by building a structure in the gaps and placing additional fill. Further laboratory and field evaluation of this concept is needed; however, it could prove to be an effective way to maintain a shoreline exposed to rising sea levels.

Groins. Groins are built perpendicular to the shore to trap sand transported alongshore by waves and/or to hold existing sand from being transported away (see Figure 6-2). Many people mistakenly call these structures jetties. They have little effect on the offshore transport of sand during storms, unless the angle of wave attack is extremely oblique. Typically, groins extend from the beach berm crest to the outer edge of the surf zone. Groins are most commonly rubble mound structures, but they have been built of concrete or wood sheet piling, concrete blocks, or timber cribs filled with stone.

Because they are perpendicular to shore and located mostly in shallower water than offshore breakwaters, the cost per unit length of structure for groins is typically less. For example, the groin extension at Lake View Park, Ohio, cost approximately $1,476/m ($450/ft) of structure (see U.S. Army Corps of Engineers, Buffalo District, 1977) compared to $2,919/m

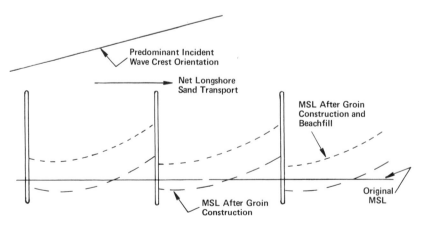

Figure 6-2. Schematic plan view—typical groin field.

($890/ft) for the offshore breakwater at the same site. Wood pile groins on the Atlantic Coast of Delaware had an estimated cost of $984/m ($300/ft) in 1975 (see U.S. Army Corps of Engineers, Philadelphia District, 1975). In a groin field, the ratio of groin length to distance between groins can vary from 1:1.5 up to 1:4. Using a typical value of 1:2 and assuming groins cost half as much as offshore breakwaters on a per foot of structure basis, typical 1980 costs per foot of shoreline for groins could vary from $100 to $1,500 depending on level of wave attack, whether or not beach fill will be placed, and beach slope.

With a significant sea level rise, existing groins would have to have their crest raised, and unless beach fill is provided to counter inundation, they would have to be extended inland. Groins would not directly control shoreline recession in response to sea level rise. However, where there is a substantial alongshore movement of sand, they can help control erosion from sea level rise.

Revetments. A revetment is a structure typically consisting of loose armor material, stones, and concrete blocks laid on a relatively flat slope to protect an embankment from wave attack (see Figure 6-3). Revetments are used in locations where there is little or no protective beach and low to moderate wave climate, such as the Chesapeake Bay. They would rarely be used on open ocean shorelines. Important considerations in their design include: a filter, finer stone, or cloth to keep embankment soil from being removed, and toe protection, sheet pile cutoff wall, or larger stone to prevent failure from scour at the toe. An adequately designed revetment is an effective means of stabilizing a shoreline subject to low-moderate

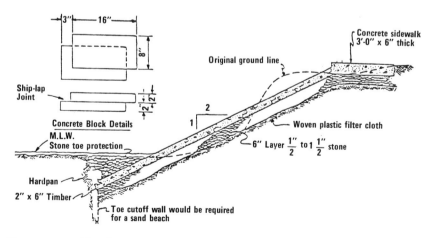

Figure 6-3. Profile of concrete block revetment showing the most basic features of a typical revetment.

Table 6-2. Cost Estimates for Bluff Protection Revetment

Armor Material	Length along Slope (ft)	Cost/Ft
Stone riprap	52	$225
Gabion mat (wire cover filled with stone)	54	$136
Lok-Gard concrete blocks	71	$201
Car tire mat filled with cement	61	$125

Source: Data from B.L. McCartney, 1976, Survey of Control Revetment Types, CERC MR 76-7, Fort Belvoir, Va.: Coastal Engineering Research Center.

waves. Fill can be placed behind the revetment to raise ground levels along the shore. A flat near-shore slope and adequate sand to satisfy alongshore transport greatly improve their performance.

McCartney (1976) estimated the costs for a bluff protection revetment on Lake Superior having a 20 year design life. The bluff slope is about 1:3, no special toe protection was required, and no overtopping by wave runup was allowed. The results for four revetment types (1975 cost figures) are presented in Table 6-2. McCartney's cost estimates are probably somewhat low, as can be seen by comparing some of his unit costs with those in other sources (see U.S. Army Corps of Engineers, Philadelphia District, 1980). Thus, in 1980 figures, these revetments might cost $984-$1,640/m ($300-$500/ft) of beach for an exposure similar to Lake Superior's. Designing for a longer project life at a site having a different embankment slope could raise the cost per meter of beach to, say, $1,640-$1,968 ($500-$600/ft).

Because of higher waves reaching the revetment, higher sea levels will require that revetment crests be raised and, at some point, armor units sizes be increased and toe scour prevention works be improved. For any given sea level rise, the structure and landward supporting embankment can be appropriately raised to continue control of erosion and inundation. Quoted structure costs do not include any costs for enlarging the embankment that supports the revetment.

Dikes. A dike or levee is an earth fill mound, usually having a trapezoidal cross-section, that is placed along the land/water edge to prevent water from flooding the lower dry land area. The side exposed to waves and currents is often revetted with rip rap, asphalt or concrete pavement, and special flexible matting. An impervious core, for example, a clay layer, is desirable to limit seepage through the dike during raised water levels. To control water overtopping the dike by wave action, it is desirable to have water collection channels and sump pumps to remove the water. Some of these features are demonstrated in Figure 6-4, which shows a typical levee section for proposed flood control works where strong wave attack is not anticipated.

When space is available (base width required is about five times the structure height) and fill can be found, a dike system is the best way to control flooding, except where there is strong exposure to wave action such as the open ocean. For a given sea level rise, greater wave action requires a more massive revetment system and a higher structure crest to control increased wave runup. Both features can significantly increase the cost of the dike. An important advantage of a dike system for controlling sea level rise is that it can be easily raised by placing additional fill on the top and backside and extending the revetment. As sea level rises and the land landward of the dike is now continuously below sea level, a canal/pump system will be needed to remove water that seeps into the area as well as normal runoff from rainfall. Lock systems would have to be constructed to connect interior navigation channels with the sea.

The cost per foot of a dike can vary widely depending on its dimensions, availability of fill and impervious core material, need for revetment of the exposed side, accessibility of the work site, length of section being constructed, and other factors. In 1980 prices, a small nonrevetted levee with the dimensions shown in Figure 6-4 might cost $492-$656/m ($150-$200/ft) of structure. A 6.09 m (20 ft) high revetted levee, on the other hand, could cost $3,280-$3,936/m ($1,000-$1,200/ft). These cost estimates are typical, but actual costs at a given site can vary substantially.

Floodwalls. Floodwalls, usually made of concrete, are used in urban areas where broad earthen structures such as dikes would use too much valuable land. The function of a floodwall is to protect a land area from flooding. As with levees, riprap is often added in front of a dike, as shown in Figure 6-5. This could be done with existing walls as sea level rises.

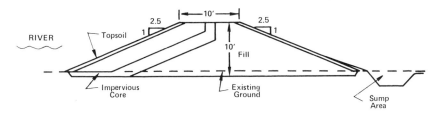

Figure 6-4. Typical dike or levee section to prevent inundation where strong wave action is not anticipated.

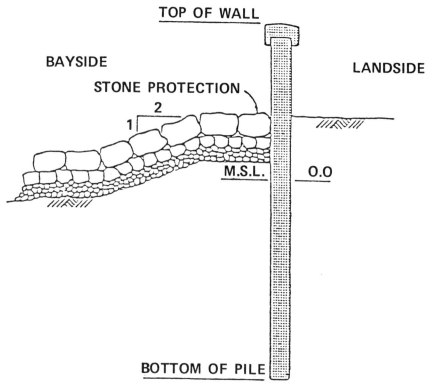

Figure 6-5. Wall and levee section.

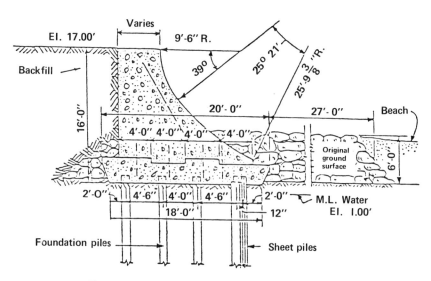

Figure 6-6. Massive concrete seawall, Galveston.

194

Seawalls.   In areas of extreme wave action where shore erosion and inundation due to sea level rise and storm surge are to be completely controlled, a concrete seawall may be constructed. Figure 6-6 shows the massive concrete seawall built in Galveston to resist storm surge and wave action. Note the sheet piling driven at the toe to resist scour at the base of the structure, bearing piles to support the structure, a curved or stepped face to limit wave overtopping, and backfill to raise the land elevation behind the structure. Seawalls require less space than dikes or levees.

A seawall is much more costly than a revetment or bulkhead and consequently would only be used in areas of strong wave attack or where valuable property is to be protected. Because of the necessary concrete form work, reinforcing steel, and bearing and support piles, seawalls cost $9,840 or more per meter ($3,000/ft) of structure.

If appropriately designed, a seawall could be extended vertically at a later time, in response to a rising sea. This would entail, for example, providing adequate bearing and cutoff piles, since deeper water would typically mean greater scour, and designing the top so a uniform strong connection can be made as the structure is raised. Also, the original section should be designed to withstand the larger and more frequent waves that will attack the structure during higher sea levels.

Seawalls have also been built at the base of erodable cliff sections to prevent cliff retreat caused by storm wave attack. Since their purpose here is erosion control, not prevention of inundation, they may be made of rubble mounds containing large breakwater-size armor units.

Bulkheads.   Typically, a bulkhead is a vertical wall constructed at the land-water interface and having the primary purpose of retaining fill (see Figure 6-7). Bulkheads are commonly found in areas where strong wave and current action are not likely, such as marinas and harbors, and along inland waterways. When they must resist wave attack, they are more massive and fronted by a beach and/or rubble toe scour protection. Bulkheads can be built from steel, aluminum, timber, or concrete sheet piling. Anchor piles or "tie-backs" are usually required to keep the bulkhead from falling into the water because of the pressures exerted by the fill the bulkheads retain.

As sea level rise and coastal inundation occur, much of the shoreline requiring protection from inundation will be in sheltered or semi-sheltered bays and estuaries and thus appropriate for protection by bulkheading. As deteriorated bulkheading is replaced and a significant sea level rise is anticipated, the replacement bulkheading length should anticipate this rise. Typical projected lives for quality timber and steel bulkheading is 25 years; concrete bulkheading might have a projected life of 50 years.

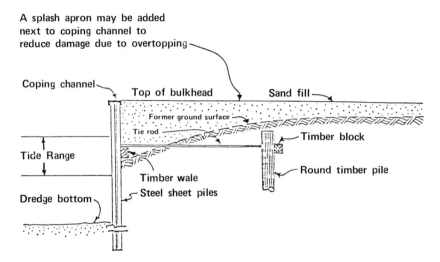

A splash apron may be added
next to coping channel to
reduce damage due to overtopping

Coping channel
Top of bulkhead
Sand fill

Former ground surface
Tie rod
Timber block

Tide Range

Timber wale
Round timber pile
Steel sheet piles

Dredge bottom

Figure 6-7. Steel sheetpile bulkhead, Nantucket Island, Massachusetts.

Where damage from vessel impact, ice, and so on is expected or when the original structure quality is not the best, shorter projected structure lives can be expected.

Steel and timber bulkhead designs and their costs for Great Lakes usage are given by the U.S. Army Corps of Engineers, North Central Division (1978). The cost per foot for bulkheads including steel piling with cable tiebacks, sandfill, and stone toe protection are found in Table 6-3. Timber bulkheading, which has a shorter design life, costs $230, $312, and $492/m ($70, $95, and $150/ft) for the same conditions.

Prestressed concrete (7 m/long, 0.3 m thick; 23 ft long, 1 ft thick), timber (8 m long, 3 m thick; 27 ft long, 10 in thick) and Z-27 steel (8 m long; 27 ft long) piling bulkheads were designed for the Atlantic coast of Delaware (see U.S. Army Corps of Engineers, Philadelphia District, 1975). They were part of a beach erosion control/hurricane flooding protection project that included groins and beach nourishment. The costs per meter (1975) were $2,162 for prestressed concrete ($659/ft), $2,775 for timber ($846/ft), and $2,060 for steel ($628/ft), plus $577/m ($176/ft) for riprap toe protection.

From the above figures, typical costs per meter for bulkheading in 1980 could be expected to vary from $492 to $984 ($150-$300/ft) in sheltered harbor/waterway areas and from $3,280 to $3,936/m ($1,000-$1,200/ft) on the open coast with a protective beach that might be removed only during heavy storms.

Table 6-3. Bulkhead Costs

Water Depth 50 Ft Offshore (ft)	Piling Length (ft)	Cost/Ft
3-4	13	$290
5-6	20	$460
7-8	25	$580

Source: Data from U.S. Army Corps of Engineers, North Central Division, 1978, "Help Yourself: A Discussion of Erosion Problems on the Great Lakes and Alternative Methods of Protection," pamphlet.

Dams. Dams or tidal barriers are concrete or earthfill structures built across an estuary or tidal river to prevent a storm surge from moving up river and causing flooding in tributaries and in the main stem. A levee system along the river or estuary and running inland is used in conjunction with the dam to prevent flanking by floodwaters. Gates are provided through the barrier to allow navigation vessels to move through and to allow drainage to the sea during normal tidal elevations. The gates must be closed at the approach of a storm and a pumping facility must then pass the river flow over the barrier.

The cost of such dams is very site specific. Some information is given by Childs (1965) for New England tidal barriers.

Artificial Beach Nourishment. Eroding shorelines can be stabilized by the placement of suitable (adequate particle size) sand, usually a large initial fill followed by periodic renourishment to make up for losses. Beach nourishment, by raising beach surface elevations, will also act to limit inundation. Many beach fills are stabilized by groins and/or offshore breakwaters to reduce renourishment requirements. In turn, beach nourishment is used to stabilize some shore protection structures, for example, bulkheads, dikes. For example, groins with beach nourishment are used at critical areas of the Dutch coast to protect the base of potentially erodable dunes. To be feasible, a good source of sand located near the nourishment area is required. Typical sources included offshore deposits, deposits at the ebb and flood deltas of a tidal inlet, and occasionally, if adequate quantities can be found, onshore or in nearshore embayments. Mechanical bypassing of sand past an obstruction to alongshore transport such as an inlet is a form of beach nourishment; sand is taken from the site that is accumulating and taken to the site that is eroding.

Numerous beach nourishment projects have been completed in the United States during the past few decades. Hobson (1977) discusses 20 of

Table 6-4. Beach Nourishment Costs

Site	Fill Period	Volume (Yd³)	Cost/Yd³
Rockaway Beach, N.Y.	1975-1977	2,145,000	$4.38
Caroline Beach, N.C.	1971	447,000	$8.73
Hunting Isle, S.C.	1968	436,000	$1.40
Presque Isle, Pa.	1980	500,000	$7.56
Lake View Park, Oh.	1977	111,000	$7.49
Lakeshore Park, Oh.	1982	36,700	$12.08

Source: Data from R. D. Hobson, 1977, *Review of Design Elements for Beach-Fill Evaluation,* CERC 77-6, Fort Belvoir, Va.: Coastal Engineering Research Center.

these including cost data. At a given date, the cost per cubic yard of sand in place on a beach can vary widely depending primarily on the volume of fill required and the transport distance to an adequate source. This is demonstrated in Table 6-4. The typical cost per cubic meter for beach fill could vary between $7 and $13 ($5 and $10/yd³), with the former figure being for large fills and nearby sources and the latter for smaller fills and more remote sources.

To determine the cost of beach nourishment per unit length of beach, one needs to know the volume placed per unit length. Typical volume per unit length values cannot be stated, as they vary too widely depending on the desired widening of the beach, the initial beach profile and the resulting stable fill profile, and whether or not frontal dunes are to be constructed with the fill.

As sea level rises, beach fill can be placed along with stablilizing structures to maintain the location of mean sea level in critical areas. The fill would prevent inundation, and structures would control erosion of the fill. Nourishing a beach to retain a pre-sea level rise location steepens the beach face, making it more prone to erosion and more in need of stabilization by structures. In many areas where beach nourishment is practiced, supplies of suitable sand are limited, and continuous extensive placement of fill on beaches to control shoreline retreat would not be practical.

Dune Building. A line of continuous coastal dunes located just landward of the active beach profile can help limit storm inundation and beach erosion. For the former, the dune field acts like a dike and for the latter, it provides a reservoir of sand to overcome the erosive effect of waves. If an inadequate dune field exists, it may be raised and widened rapidly by the mechanical placement of sand or more slowly by trapping wind-blown sand with fences and/or vegetation. Vegetation is generally more appropriate.

The cost of the mechanical building of dunes can be estimated from the volume of sand required and the costs per cubic meter reported above. For example, a dune built to 3 m (10 ft) above a given elevation may have 14 m^3 of sand per meter of beach (500 ft^3 or about 18 yd^3 of sand per ft). At $10/m^3 ($8/yd^3), dunes would cost $472/m ($144/ft) of beach plus the costs of stablilizing the dune. The latter would include planting of vegetation and periodic chemical fertilization. In some circumstances, it may pay to revet the seaward face of a natural or artificial dune to increase its resistance to erosion by storm waves.

Marsh Building. In salty or brackish estuarine areas, salt marshes are common. A shallow flat marsh can provide some protection to adjacent land areas by dissipating incident wind and vessel waves. Also, vegetation in the marsh encourages trapping and stablilization of fine sediments and upward growth of the marsh surface (at quite a slow rate). If insufficient vegetation exists, marsh growth and stabilization can be encouraged by the planting of vegetation. Woodhouse (1979) presents the techniques for building salt marshes with vegetation, and Knutson and Inskeep (1982) discuss the use of salt marsh vegetation for controlling shore erosion in sheltered coastal areas.

Depending on marsh width, plant density, soil characteristics, and shoreline geometry, marshes can be stabilized against wave attack if wave generation fetches are typically a few miles or less. Planting of a 9 m (30 ft) wide strip of marsh would cost $16-$33/m ($5-$10/ft) in 1980.

With a sufficiently slow sea level rise, marsh growth may keep pace with increasing sea levels and resist erosion/inundation effects. However, significant inundation and related higher waves would soon destroy a vegetated marsh.

SALTWATER INTRUSION

The Process of Saltwater Intrusion

Many investigations have been conducted to determine the movement and extent of saltwater intrusion. A brief summary of the principles is presented here.

Saltwater Intrusion Into Aquifers. The Ghyben-Herzberg principle provides an initial estimate of the inland extent of saltwater intrusion in a simple unconfined aquifer of infinite depth (see Figure 6-8). This theory assumes two fluids separated by a sharp interface and ignores many of the complexities found in real aquifers (see Figure 6-9). The principle

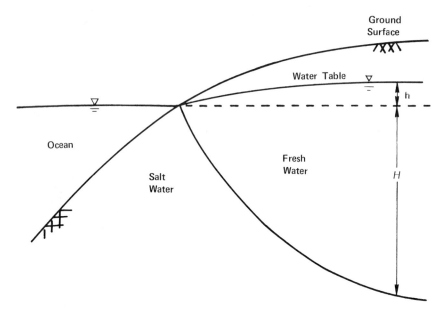

Figure 6-8. Saltwater intrusion in a coastal aquifer according to d'Andrimont.

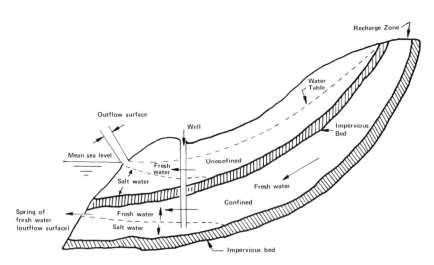

Figure 6-9. Saltwater intrusion into two-aquifer system. *(After J.S. Brown, 1925, Study of Coastal Groundwater With Special Reference to Connecticut, U.S. Geological Survey Water Supply Paper 537.)*

assumes that an equilibrium condition exists between the saltwater offshore and a freshwater flowing from the upland area down toward the ocean. As shown in Figure 6-8, because the saltwater is 1.025 times denser than the freshwater, the saltwater/freshwater interface lies a distance *below* mean sea level (H) for a given height of the freshwater *above* mean sea level (h). The product of the density of saltwater times its height is balanced by the density of freshwater times its height. In equation form:

$$1.025 \times H = 1.0 \times (h + H)$$

or solving for H, the interface location below mean sea level in terms of h, the freshwater head above mean sea level is $H = 40\ h$. At any point in time, for every foot that the freshwater table lies above mean sea level, the depth to the saltwater is 12 m (40 ft) below mean sea level.

Where coastal aquifers are strongly influenced by the withdrawal of water, the location of the saltwater front is controlled by the pumping pattern and intensity rather than the density balance predicted by the simplistic Ghyben-Herzberg principle (see Figure 6-8). Aquifers with complex geometries or heavy groundwater pumpage may require the application of models that are more sophisticated than the Ghyben-Herzberg model, as addressed by investigators such as Pinder and Gray (1977) and Contractor (1980). The mixing or diffusion zone between the saltwater and freshwater is taken into account in many of these models. However, it is often relatively small, and for an initial study a relatively simple model is recommended that assumes that a sharp interface exists between the freshwater and saltwater.

Fortunately, moderately sophisticated models, which are not over-simplified and do not require an unrealistic effort to apply, exist. Many investigators such as Harbaugh et al. (1980) have employed such models to predict the movement of the saltwater front in aquifers for various pumping schemes without resorting to the extremely complicated convection-dispersion transport models. Existing saline water levels obtained by field measurements are converted to equivalent freshwater levels. Along with saltwater velocities computed by the model, useful initial estimates of saltwater movement can be obtained. However, these methods will not provide detailed description of the location of the saltwater equilibrium surface. In many instances, the choice of the proper model is crucial.

The following initial data are required to employ such models properly: an existing water level map, an existing chlorine concentration map based upon field data, the physical properties of the aquifer, and current, past, and projected groundwater withdrawals. Then, a predictive model should be chosen which can model the projected impact of the rise in sea level.

Saltwater Intrusion into Estuaries. During extended droughts, decreased river flow allows the saline water to migrate up the estuary. A rise in sea level will also cause saltwater to migrate upstream. The general methods of preventing saltwater intrusion up estuaries are similar for sea level rise, drought conditions, and storm surge. As discussed previously, storm surge elevates the ocean in relation to the estuary water level, causing saltwater intrusion. A major difference is that storm surge and drought conditions last for a limited duration, whereas the sea level rise is expected to last much longer.

In order to minimize saltwater migration, river basin commissions provide low-flow augmentation and water conservation requirements during periods of low flow. Water from rainfall and snowmelt are stored in large surface reservoirs and released continuously during droughts to maintain a flow that helps repel the saltwater from migrating upstream. These planning agencies recognize the need to sustain stream flows to protect freshwater intakes, instream uses (including fish migration and fish production), and shellfish beds, as well as treated-waste assimilation, recreation, and salinity repulsion. An economic justification is usually necessary, showing that the cost of the mitigation is less than the anticipated benefits.

The prevention of saltwater intrusion can be provided by other options including:

Barriers. Dams can be constructed that physically prevent the saltwater from moving past a certain point in the estuary. Injection barriers have also been employed successfully.

Restrictions on pathways for saltwater intrusion. Construction of canals allow saltwater to migrate into inland areas and allow a pathway for saltwater intrusion to occur.

Alternate sources of water. Water users may be able to obtain water from other sources that are not endangered by saltwater intrusion.

Restrictions on use of water. During periods of higher sea level or drought, stricter conservation and restrictions on export of water from the river basin may be considered for short durations.

The Delaware River Basin Commission (DRBC) is one of the planning agencies responsible for managing the surface waters and groundwaters within the drainage basin of the Delaware River. The DRBC has included sea level rise projections through 2000 in its planning (DRBC, 1981) and is currently working on a cooperative program with Dr. Gerard Lennon of

Lehigh University and the Environmental Protection Agency in estimating the salinity intrusion into the Delaware Estuary for the sea level rise scenarios discussed in Chapter 3.

Control of Saltwater Intrusion

Control methods for saltwater intrusion have been employed or seriously considered only in areas where withdrawals of water have caused water levels in aquifers to fall significantly below mean sea level. Because of the very slow velocity with which the saltwater moves, many localities with serious overdrafts have not yet lost their aquifers as sources of water. However, they must solve this problem eventually because once saltwater has invaded an aquifer, it could take hundreds of years to regain the salinity levels of the virgin aquifer.

Where the existing water levels in principal aquifers are already several tens of meters below sea level, a rise in sea level of less than 1 m would be of less consequence than a slight increase in the withdrawal rate. However, in areas where the existing water levels are within a few meters of mean sea level, the impact could be significant. If sea level rises more than 1 m, all coastal aquifers will be affected to some degree.

The greatest danger to freshwater aquifer supplies could be the migration of saltwater up an estuary that recharges an aquifer. If the water levels in the aquifer are below mean sea level because of withdrawals, the saltwater would recharge the aquifer.

Several control strategies can be used to prevent or retard saltwater intrusion into aquifers. They include:

Physical subsurface barriers. Options include driving sheet pile, installing a clay trench, or injecting impermeable materials through wells.
Extraction barriers. The saltwater that moves inland is collected and removed. The pumping encourages further intrusion and may inadvertently withdraw freshwater.
Freshwater injection barriers. Freshwater from another source is injected into the aquifer, raising water levels in the area and reversing the saltwater intrusion.
Increased recharge. Spreading of water on the land in upland recharge areas allows more percolation (infiltration of water into the aquifer), which retards saltwater intrusion.
Modified pumping patterns. Reducing withdrawals or moving the pumping locations further inland can substantially reduce the intrusion.
Direct surface delivery to replace groundwater use. Groundwater can be replaced by surface water through the use of direct surface delivery.

Combinations of these techniques can also be employed. A combination of an extraction and an injection barrier or increased recharge with an injection barrier are particularly effective combinations.

Physical Barriers. Subsurface physical barriers such as sheet pile cutoff walls, clay slurry trenches under earth dams, and impermeable clay walls are routinely used by engineers in the field to control the movement of water and other liquids including the containment of hazardous waste materials. It is also possible to inject materials that form a zone of low permeability. Figure 6-10 illustrates a cross-section of a typical physical barrier.

Kashef (1977) indicates that, although the construction methods are technically well established, the cost is usually too high because the required depths are substantial. Even in the uppermost layers where the cost may not be prohibitive, Kashef points out that the backwater effect could cause coastal lowlands to become waterlogged. Unit cost estimates for slurry walls range between $20 to $40 per square meter of surface area. Thus, for a wall as wide as the standard trenching equipment and 10 m deep, the cost is $200-$400 per linear meter of wall. The cost is highly dependent on depth of cutoff, length of wall, and specific material availability costs. Barriers require complete depth of cutoff to be effective.

Impermeable walls can be almost 100 percent effective at preventing saltwater intrusion. However, in actual practice, some limited penetration will occur.

Extraction Barriers. Extraction barriers have been used in various locations in order to prevent or reduce saltwater intrusion. In 1965 a 0.5 mi (0.8 km) long extraction barrier was employed in the Oxnard aquifer, Oxnard Plain, Ventura County by the California Department of Water Resources (CDWR), as summarized by Stone (1978). The five-well experimental extraction barrier was discontinued in 1968 because of corrosion and proved to be inadequate at preventing the intrusion. Figure 6-11 illustrates a typical extraction type barrier where the salt-water intrusion is halted by the withdrawal of saltwater relatively close to the shoreline.

Extraction barriers may withdraw some freshwater that would otherwise be useful and thus may not be a valuable option where water supplies are scarce. In addition, problems with saltwater corrosion must be overcome.

Again, experiences by Kashef (1977), Stone (1978), and others have generally indicated that the saltwater intrusion caused by pumping over-drafts can be technically controlled by extraction barriers but are usually more expensive than injection barriers.

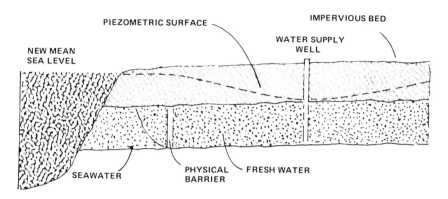

Figure 6-10. Physical seawater intrusion barrier. *(After Wayne L. Stone, 1978, "An Assessment of Alternate Sea Water Intrusion Control Strategies for the Oxnard Plain of Ventura County, California," report submitted in partial satisfaction of the requirements for the degree of Doctor of Environmental Science and Engineering, Berkeley: University of California.)*

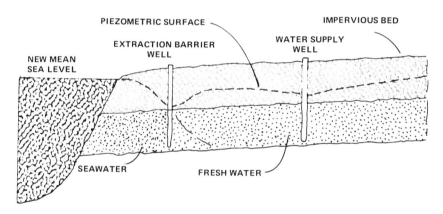

Figure 6-11. Extraction-type seawater intrusion barrier. *(After Wayne L. Stone, 1978, "An Assessment of Alternate Sea Water Intrusion Control Strategies for the Oxnard Plain of Ventura County, California," report submitted in partial satisfaction of the requirements for the degree of Doctor of Environmental Science and Engineering, Berkeley: University of California.)*

Although extraction barriers have not proven to be economically justifiable for saltwater intrusion in most localities that have considered them, certain special considerations may result in the economical use of extraction barriers. Such sites might include the prevention of saltwater intrusion into a limited area such as a hazardous waste site or a coastal aquifer with a relatively narrow connection to the ocean.

However, a major problem with the extraction barriers are that the withdrawal of saltwater and the inadvertent withdrawal of some freshwater cause the water levels to fall substantially throughout the basin. The increased lift and the cost of wells going dry often become costly in time. Furthermore, although a complete cutoff extraction barrier does not have to be completed all along the coast, saltwater intrusion can occur around the barrier. The lower levels also encourage saline water from above or below to move vertically into the aquifer. For a 1-3 m sea level rise scenario over the next 120 years, extraction barriers can be up to 100 percent effective along the length of coast being protected. However, vertical leakage may occur from above or below.

Freshwater Injection Barriers. Figure 6-12 illustrates a typical injection barrier in operation to control the saltwater intrusion for cases where the sea level is in excess of freshwater levels. In contrast to the extraction barrier, with an injection barrier, freshwater is injected into the aquifer through a line of wells along the coastline. The higher groundwater levels along the injection barrier prevent saltwater intrusion from occurring. A proper design of well spacing and location must be performed to ensure that saltwater does not intrude around the injection barrier, in between individual wells, or move vertically from above or below.

The problems with injection wells include the fact that a relatively large number of wells is required, a high maintenance cost will be necessary

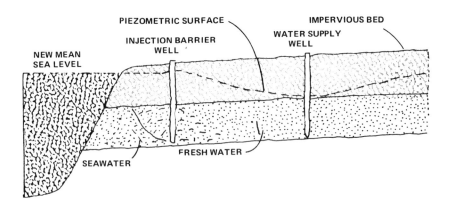

Figure 6-12. Injection-type seawater intrusion barrier. *(After Wayne L. Stone, 1978, "An Assessment of Alternate Sea Water Intrusion Control Strategies for the Oxnard Plain of Ventura County, California," report submitted in partial satisfaction of the requirements for the degree of Doctor of Environmental Science and Engineering, Berkeley: University of California.)*

to prevent plugging of wells, and most important, a source of freshwater will be needed.

The Los Angeles County Flood District injection barrier construction costs were approximately $20 million from 1953 to 1973, not including the cost of purchasing the water to be injected and not adjusting for inflation (DRBC, 1981). The annual maintenance and operation cost of the 32 km (20 mi) barrier was approximately $1.5 million in the period 1978-1980, and the annual cost of filtered injection water was $5,684,375 ($0.086/m^3 or $106.25/acre-foot—see Bookman-Edmonston Engineers, 1982). An acre-foot is a unit of water equal to 1,234 m^3 (326,000 gal).

The 1978-1980 operation averaged $44,000 per km ($70,000/mi); figuring 6 wells/km (10 wells/mi), this is $7,000 per well for an estimated 3.8 × 10^5 m^3 (100 million gal) per day of injection.

The operation cost depends upon the length of the barrier, the geometry and physical properties of the aquifer, differences in water levels in the aquifer relative to mean sea level, and the volumes of water injected, being recharged to the aquifer, and being withdrawn.

Stone (1978) summarized the capital costs ($18.3 million) and annual operation costs in 1980 ($430,000) for the Oxnard Plain Study. For the 13 km (8 mi) barrier, the 1980 capital cost is $1.4 million/km ($2.3 million/mi) and the annual maintenance cost is $34,000/km ($54,000/mi).

Also, because 1.5 × 10^7 m^3 (12,100 acre-feet) of water are to be injected annually, expressing unit costs in terms of volume of injection water rather than miles protected may be useful where the injection is to prevent intrusion due to a rise in sea level of 1 or 2 m. The unit capital cost per acre-foot of water injected annually is $1,512, with an operational cost of $0.028/m^3 ($35/acre-foot) injected.

Because the injection of water raises the water levels in the vicinity of the barrier, a complete cutoff all along the coast is not required. If the cutoff barrier is maintained at 1 or 2 m above mean sea level (continuously increasing the barrier water level as sea level increases), the injection barrier will provide 100 percent effectiveness in preventing saltwater intrusion along its length. In addition, the freshwater mound will tend to flow inland toward the lower water levels there. Also, freshwater from the injection barrier will flow along the coast for a limited distance, extending the effective length of barrier slightly. The effectiveness of the injection barrier will be maintained if the water levels in the vicinity of the barrier are increased as sea level increases to always maintain them at 1 m (3 ft) above mean sea level.

Increased Recharge. In many coastal locations in the United States, sufficient amounts of freshwater are available for recharge during periods of high precipitation. Although some water is captured during these periods and stored in surface reservoirs, very little water is recharged to groundwater reservoirs for use in drought periods. This extra water which is "wasted" to the ocean could be used to replenish the aquifer, build up groundwater levels, and repel the saltwater intrusion. If the natural plus additional recharge exceeds the groundwater withdrawals, a stable saltwater line would be established.

In many instances, such as Oxnard Plain in California (Stone, 1978), the recharge region of the principal water supply aquifer is far away from the coast. In these regions, it is possible to recharge the confined aquifer far from the shoreline and prevent saltwater intrusion. For unconfined aquifers, the recharge occurs in an area near the coastline and near the center of withdrawal.

The problems with increased recharge can be a lack of sufficient replenishment water, lack of inexpensive land for the recharge basins or shallow injection wells, and costly technical problems of maintaining an adequate inflow rate. However, as mentioned previously, many areas have excess water during wet periods, which can be utilized during dry periods.

In the Oxnard Plain, the capital cost of the replenishment water is $14 million and annual operation costs (1980 dollars) are $64,000, as summarized by Stone (1978). In addition, the cost of purchasing 31 million m^3 (25,000 acre-feet) of water annually at $0.93/m^3 ($115/acre-foot) is $2,875,000/year. Considering that the recharge is used in place of a 13 km (8 mi) injection barrier, the unit capital cost is $1.0 million/km ($1.8 million/mi) and $5,000/km ($8,000/mi) of annual operational costs, exclusive of the cost to purchase water. However, the amount of required injection water is very dependent upon the withdrawal of groundwater by users, natural recharge, and basin geometry. A more appropriate unit cost figure is the cost of recharge per unit volume of water. In this case, the capital cost is $0.45/m^3 ($560/acre-ft) injected annually plus $0.002/m^3 ($2.56/acre-foot) of water recharged.

The saltwater will intrude farther inland than is now occurring unless the amount of additional recharge can push the saltwater equilibrium surface seaward. Hence, significant saltwater intrusion can still occur even with increased recharge for a 3 m sea level scenario. A 1 m sea level rise will probably intrude only slightly with the increased recharge option.

Modified Pumping Patterns. For unconfined aquifers where no pumping exists, an intrusion of saltwater as a result of sea level rise could damage agricultural crops. Either the injection barrier or increased recharge would be viable solutions if substantial crop damage was expected without control of the saltwater advance.

For unconfined or confined aquifers where moderate pumpage already occurs and the effect of a sea level rise is projected to be important, a phased shutdown of wells can be designed as the monitored saltwater intrusion progresses. Instead of a disorganized search for alternate water as the chloride concentrations increase, logical permitting of new wells or new economical surface distribution schemes can be implemented.

The cost of new surface distribution schemes may be expensive, as is the abandonment of old wells that are still operational. Cost figures are dependent upon the existing facilities, resulting in site costs that are of little use at other locations. Cost figures prepared by Stone (1978) are very specific to the particular setting in the Oxnard Plain and are presented because costs were available for some of the previous options, thus making them comparable. Pumping plan C, the least costly of the three modified pumping schemes presented by Stone, has a capital cost of $10 million with an annual 1980 operational cost of $180,000. It is difficult to express a unit cost for the case of modified pumping plans.

A modification of pumping patterns will allow water levels to recover in critical areas. This will have an effect of slowing down the saltwater advance. However, the saltwater will intrude until it reaches a new equilibrium. Depending upon the recharge rate, the pumping rates, the net overdraft, water levels, aquifer geometry, aquifer characteristics, and the present status of saltwater intrusion, the effectiveness of the modification of pumping patterns will vary. For a saltwater interface currently in equilibrium, an increase in sea level rise may be counteracted by a modification of pumping patterns. However, it is possible that the intrusion will be retarded, not stopped. A 3 m sea level rise over the next 120 years will allow the intrusion to occur at a faster rate and the equilibrium position will be further inland in general than would occur in a lesser sea level rise scenario.

Direct Surface Delivery. Another method that can be used to prevent saltwater intrusion is direct surface water delivery in lieu of groundwater withdrawal. If a long-term but inevitable sea level rise faces the United States, the gradual phasing out of certain pumpage could be conducted in a very rational manner with the resulting demand for water satisfied by direct surface delivery. The state of New Jersey has approved a bond issue to provide hundreds of millions of dollars to study and improve the distribution of water in the state. If the state managers recognize and properly address the possibility of extreme sea level rises, the protection of water resources and control of saltwater intrusion may be possible at a very reasonable cost, provided they respond to the need in the near future. Unnecessary expense will be incurred if a new freshwater intake is constructed to withdraw freshwater from a river that will be excessively saline during the life of the intake because of saltwater encroachment as a result of a moderate sea level rise in the next 120 years.

The work by Stone (1978) indicates that an initial capital cost of $11.4 million and an annual 1980 operational cost of $70,800 will be required to prevent the saltwater intrusion in the Oxnard Plain. The capital cost per annual cubic meter delivered is $0.76 ($942/acre-foot). Other costs include a 1980 annual unit operational cost of $0.051/m^3 ($64/acre-foot) and a cost of $0.093/m^3 ($115/acre-foot).

Direct surface delivery allows less groundwater to be withdrawn, which in turn allows water levels in the aquifer to recover. The higher water levels in the aquifer will retard any existing saltwater advances, and in some instances may push the saltwater back. For an aquifer in an equilibrium situation in the face of a 1-3 m sea level rise over the next 120 years, a certain amount of the groundwater withdrawal should be replaced by direct surface delivery in order to maintain equilibrium. For a 3 m sea level rise, more drastic cutbacks would be required to keep the saltwater from intruding. Under any circumstance, 100 percent effectiveness in preventing saltwater intrusion should not be expected.

SUMMARY

This chapter has presented an overview of coastal engineering methods for controlling the effects of sea level rise. That is, hard and soft structural methods were the focus, rather than political methods. The specific effects considered were shore erosion and inundation, increased storm surge flooding, and salinity intrusion (particularly into groundwater supplies). For each of these effects, the process involved was discussed in general terms and the basic approaches to controlling the effect were discussed. Then, the specific methods that define each approach were described, including a general explanation, situations where the method can be used, typical cost data, and effectiveness of the method in controlling the effects of sea level rise.

With an accelerated sea level rise, the importance of having the best possible forecasts of the expected rate of rise cannot be understated. With this information, planning can be more effectively accomplished by setting aside space for future control works. Control works can be built to the ultimate required size and at the optimum location for sea levels that would occur during their design life (or they can be designed for easy expansion as sea levels rise), and new development that would be flooded or destroyed by erosion can be limited or prevented.

If the rise in sea level during the next century is on the order of magnitude suggested by the high rise scenarios, it is likely that eventually some form of national or state programs will be developed to respond to the crisis. Presumably, any such programs would efficiently and effectively combine political and structural responses. Precise forecasts of future sea levels based on sound scientific analysis would hasten the development of these programs. However, if a much lower but still significant sea level rise occurs, the response will more likely be on a site-by-site, ad hoc basis. This will be particularly true if precise forecasts are not developed. As structures reach the end of their effective life or are in general need of repair, they will be rebuilt or improved to respond to existing (or slightly higher) sea levels. Some marginal coastal developments and some groundwater sources will be abandoned. Developing lowland areas will only have protective works adequate for a short time period. In general, it will be more difficult to develop and operate a coordinated response to rising sea levels.

For a given sea level rise scenario, the type, extent, and cost of structural responses to the rise are extremely site dependent. Particularly important factors are the local tide range and exposure to wave action, the space available, foundation conditions, the nature of existing structures, the length of shoreline to be protected by any particular method, and local construction experience and material availability. In addition to the construction of erosion, inundation, and salinity intrusion control works, existing coastal features such as jetties, piers, marinas, port facilities, bridges, and causeways would have to be modified.

The hard and soft structural responses can vary in effectiveness from location to location. However, if sufficient design information on environmental, foundation, and related conditions is available and if sufficient funding is available for construction, in all but rare situations, an effective structural response can be built. Required concepts and construction techniques to respond structurally to sea level rise are within the state-of-the-art.

In order to develop a meaningful estimate of the cost and physical extent of effort required for the structural response to a given sea level rise scenario, the following investigation is recommended. A well-developed specific estuary/coastal location should be selected where erosion, inundation, storm surge flooding, and salinity intrusion (estuarial and groundwater) are existing or potential problems. An example would be the Raritan Bay in New Jersey and the adjacent shorelines from Sandy Hook to Asbury Park. This region includes coastal urban development, port facilities, small marinas, natural unstabilized shorelines, shorelines strongly fortified by existing structures, heavy marine commerce, and so on. For a particular sea level rise scenario, design tide and storm levels would be determined, design wave climates (with appropriate return periods for the types and locations of structures to be considered) would be forecasted, estuary/groundwater salinity changes would be estimated, and potential shoreline erosion/deposition changes would be evaluated. Expected patterns of growth during the duration of the sea level rise scenario would also be projected. Then, the estuary and coastal works required to control the effects of sea level rise would be located, designed, and evaluated for cost, including required modifications to existing structures. Designs would be preliminary in nature, but they would account for local wave, water level, foundation, and layout conditions, the time during the sea level rise when the work was required, and the conditions of existing structures. It is only in this way that the precision required for accurate estimates of the potential costs of a structural response to sea level rise can be developed.

NOTES

1. The material presented herein was compiled largely from local (Philadelphia and New York) U.S. Army Corps of Engineers district libraries, the U.S. Geological Survey and the Delaware River Basin Commission offices in Trenton, New Jersey, and the Water Resources Archives Library of the University of California at Berkeley. Time, funding, and space constraints prohibit a more thorough survey of the available literature, particularly in regard to examples of the cost and effectiveness of specific control methods.
2. The U.S. Army Coastal Engineering Research Center (1977) provides a general discussion of the structural and functional aspects of these methods.

REFERENCES

Bookman-Edmonston Engineering. 1982. "Annual Survey Report on Ground-water Replenishment, 1982." Prepared for the Central and West Basin Water Replenishment District, Calif.

Brown, J. S. 1925. *Study of Coastal Groundwater with Special Reference to Connecticut.* U.S. Geological Survey Water Supply Paper 537.

Childs, E. F. 1965. *Operation of Hurricane Barriers in New England.* Fort Belvoir, Va.: Coastal Engineering Research Center, chap. 42.

Contractor, D. N. 1980. *A Review of Techniques for Studying Freshwater Salt Water Relationships in Coastal and Island Groundwater Flow Systems.* Technical report 11, OWRT Project A-00-7. Manila: University of Guam, Water Resources Research Center.

Delaware River Basin Commission. 1981. "The Final Report and Environmental Impact Statement of the Level B. Study." West Trenton, N.J.

Dunham, J. W. 1968. "Proposed Santa Monica Causeway Project." *Journal of the Waterways and Harbors Division* 94(WW1):425-436.

Harbaugh, A. W., J. E. Luzier, and F. Stellerine. 1980. *Computer-Model Analysis of the Use of Delaware River Water to Supplement Water from the Potomac-*

Raritan-Magothy Aquifer System in Southern New Jersey. Del.: U.S. Geological Survey Water Resources Investigation 80-31.

Hobson, R. D. 1977. *Review of Design Elements for Beach-Fill Evaluation.* CERC 77-6. Fort Belvoir, Va.: Coastal Engineering Research Center.

Ippen, A. T. 1966. *Estuary and Coastline Hydrodynamics.* N.Y.: McGraw-Hill.

Kashef, A. I. 1977. *Management and Control of Salt Water Intrusion in Coastal Aquifers: Critical Reviews in Environmental Control.* Boca Raton, Fla.: CRC Press.

Knutson, P. L., and M. R. Inskeep. 1982. *Shore Erosion Control with Salt Marsh Vegetation.* CETA 82-3. Fort Belvoir, Va: Coastal Engineering Research Center.

Komar, P. D. 1976. *Beach Processes and Sedimentation.* New York: Prentice-Hall, pp. 160-166.

McCartney, B.L. 1976. *Survey of Coastal Revetment Types.* CERC MR 76-7. Fort Belvoir, Va.: Coastal Engineering Research Center.

Pinder, G. F., and W. G. Gray. 1977. *Finite Element Simulation in Surface and Subsurface Hydrology.* New York: Academic Press.

Sorensen, R. M. 1978. *Basic Coastal Engineering.* New York: John Wiley & Sons.

Stone, Wayne L. 1978. "An Assessment of Alternate Seawater Intrusion Control Strategies for the Oxnard Plain of Ventura County, California." Report submitted in partial satisfaction of the requirements for the degree of Doctor of Environmental Science and Engineering. Berkeley: University of California.

Todd, D. K. 1980. *Groundwater Hydrology.* New York: John Wiley & Sons.

U.S. Army Coastal Engineering Research Center. 1977. *Shore Protection Manual.* Ft. Belvoir, Va.: Coastal Engineering Research Center.

U.S. Army Corps of Engineers, Buffalo District. 1977. "Construction Contract Cost Bidding Schedule: Beach Erosion Control Project, Lake View Park, Lorain, Ohio."

U.S. Army Corps of Engineers, Buffalo District. 1980. "Construction Cost Estimate: Beach Erosion Control Project, Presque Isle, Erie, Pennsylvania."

U.S. Army Corps of Engineers, Buffalo District. 1982. "Construction Contract Cost Bidding Schedule: Beach Erosion Control and Shoreline Protection Project, Lakeshore Park, Ashtabula, Ohio."

U.S. Army Corps of Engineers, Galveston District. 1979. "Texas Coast Hurricane Study."

U.S. Army Corps of Engineers, New England District. n.d. "Hurricane Protection Project, New Bedford-Fairhaven Barrier, New Bedford Harbor, Massachusetts." GDM 1 & 3.

U.S. Army Corps of Engineers, North Central Division. 1978. "Help Yourself: A Discussion of Erosion Problems on the Great Lakes and Alternative Methods of Shore Protection." pamphlet.

U.S. Army Corps of Engineers, Philadelphia District. 1975. "Beach Erosion Control and Hurricane Protection: Delaware Coast." General Design Memorandum Phase 2.

U.S. Army Corps of Engineers, Philadelphia District. 1978a. "Slaughter Beach, Delaware." Pre-construction report, Shoreline Erosion Control Demonstration Program.

U.S. Army Corps of Engineers, Philadelphia District. 1978b. "Kitts Hummock, Delaware." Pre-construction report, Shoreline Erosion Control Demonstration Program.

U.S. Army Corps of Engineers, Philadelphia District. 1980. "Mitigation of Erosion Damage, Delaware River, Pennsville, New Jersey." Abbreviated detailed project report.

U.S. Army Corps of Engineers, Philadelphia District. 1981. "Barnegat Inlet New Jersey." Phase 1 general design memorandum.

Woodhouse, W. W. 1979. *Building Salt Marshes Along the Coasts of the Continental United States.* CERC SR-4. Fort Belvoir, Va.: Coastal Engineering Research Center.

BIBLIOGRAPHY

Cooper, H. H., Jr., F. A. Kohout, H. R. Henry, and R. E. Glover. 1965. *Sea Water in Coastal Aquifers.* U.S. Geological Survey Water Supply Paper 1613.

Delaware River Basin Commission. 1982. "Groundwater Management Plan for Study Area 1 Coastal Plain Formations." Final report. West Trenton, N.J.: Delaware River Basin Commission.

Freeze, R. A., and J. A. Cherry. 1979. *Groundwater.* Englewood Cliffs, N.J.: Prentice-Hall.

Hull, C. H. J. 1979. *Sea-Level Trend and Salinity in the Delaware Estuary.* Internal staff paper. West Trenton, N.J.: Delaware River Basin Commission.

Hull, C. H. J., and R. Tortoriello. 1980. *Delaware Salinity Modeling Study: Effects of Simulation-Period Duration on Maximum Salinity Levels Attained.* Internal staff paper. West Trenton, N.J.: Delaware River Basin Commission.

Luzier, J. E. 1980. *Digital-Simulation and Projection of Head Changes in the Potomac-Raritan-Magothy Aquifer System, Coastal Plain, New Jersey.* U.S. Geological Survey Water Resources Investigations, 80-11.

McIlwain, R. R., W. T. Pitts, and C. C. Evans. 1970. "West Coast Basin Barrier Project, 1967-1969." Report by the Los Angeles County Flood Control District on the Control of Seawater Intrusion. Los Angeles, Ca.

Segol, G., and G. F. Pinder. 1974. "INTRUSION—Simulates Salt Water Intrusion in Coastal Aquifers Including the Effects of Dispersion." Computer program available on a limited basis from Water Resources Program, Princeton, N.J.

Segol, G., and G. F. Pinder. 1976. "Transient Simulation of Saltwater Intrusion in Southeastern Florida." *Water Resources Research* 12(1):65-70.

Stewart, M. T. 1982. "Evaluation of Electromagnetic Methods for Rapid Mapping of Salt Water Interfaces in Coastal Aquifers." *Groundwater* 20(5):538-545.

U.S. Army Corps of Engineers, New England District. n.d. "Fox Point Hurricane Barrier, Providence River, Rhode Island." GDM 4 & 5.

U.S. Army Corps of Engineers, Philadelphia District. 1979. "Delaware Estuary Salinity Study, Eight Year-Long Calculations Using the Modified Transient Salinity Intrusion Program." Prepared by M. L. Thatcher.

U.S. Army Corps of Engineers, Philadelphia District. 1983. "Delaware Estuary Salinity Intrusion Study."

Chapter 7

Economic Analysis of Sea Level Rise: Methods and Results

Michael J. Gibbs

INTRODUCTION

The direct physical effects of sea level rise will have a major influence on the use of the coastal zone throughout the country. An examination of these physical effects is but a first step in estimating the impacts of sea level rise on coastal communities and society. The importance of these impacts will depend on how we prepare for them.

Given our current understanding of the potential for future sea level rise and the opportunities to improve our understanding, we should identify the course of action that would best prepare us for the future. The choice of which actions to take (such as increasing research, constructing protective structures, or altering development patterns) requires balancing uncertain risks and costs.

Because many of the actions to prepare for future sea level rise must be taken collectively, extensive analysis and political debate on the relative importance of the risks and costs should precede decisions of whether to undertake certain actions. Additionally, individuals must decide for themselves whether the potential for future sea level rise should alter their current and future private activities (such as purchasing ocean-front property).

The objective of the project summarized here is to estimate what is at stake in these public and private decisions. Methods were developed and implemented to answer two questions. First, if we take no special actions to prepare for sea level rise, what is the impact on society if it in fact occurs? And second, by how much can we reduce the impact of sea level rise if we take actions to prepare for it? If the impact is large but we can reduce it substantially through preparation, then the decision regarding how best to prepare is an important one. If the impact is small or if preparation has little benefit, then the decision is not so important.

The analyses and results presented below conclude that both the impacts of sea level rise and the value of preparation are large indeed. Based on the analyses of the physical impacts of sea level rise presented in the previous chapters, the economic impact of sea level rise on Charleston and Galveston is estimated to be hundreds of millions, perhaps billions, of dollars. Preparing for future sea level rise could reduce these impacts by over 60 percent in some cases. It appears, therefore, that the stakes are high.

Like the other parts of the project described in this book, the analysis presented in this chapter is a first attempt to examine a relatively unstudied phenomenon. The analysis presented here must be refined and extended in a variety of ways. The estimated impacts of sea level rise reported below are conservative because quantitative estimates could not be made for several effects and the set of preparation actions considered is limited. More refined analyses of selected individual and public actions would improve the precision of the estimates. Nevertheless, the results serve as a first step toward a better understanding of the potential economic and societal impacts of sea level rise.

ANALYTIC METHODS

The methods developed for this analysis are based on the principles of welfare economics. The two quantities investigated, the impact of sea level rise and the value of anticipating sea level rise and preparing for it, were measured in terms of the net economic cost from the viewpoint of a community or study area. As is generally the case with economic analyses, distributional impacts are not valued; that is, if one person gains $10 and another loses $10, the net impact is estimated as zero, with no value placed on the change in the distribution of the $10. Consequently, the distributional and equity implications of sea level rise are not discussed.

This section is divided into two parts: an analysis of economic impacts and an analysis of the value of anticipating sea level rise. Before describing the details of the methods used, the following brief example provides an intuitive feel for the two quantities estimated. The example concerns a hypothetical Community X which, under alternative assumptions, undertakes three sets of economic activities: A, B, and C. The basic approach described in this example is applied below to the Charleston and Galveston study areas.

Community X is a moderate-sized coastal city. Being located on the coast, parts of the city experience erosion and run the risk of being damaged by storms and flooding. The variance of the erosion and storm hazards throughout the city is reflected by existing zoning and development patterns.[1]

If the sea level does not rise over the next 100 years, Community X will carry on a particular set of economic activities; call this Set A. These activities may include manufacturing (a refinery), transportation services (a port), housing for its inhabitants, and tourist and recreation services. Set A may include purchasing goods from other areas (such as raw materials) or supplying goods to other areas (such as finished products). This set of activities will have some economic value, which is called the net economic service value.

If the sea level does rise, Community X may carry on a slightly different set of economic activities; call this Set B. Set B may differ from Set A because areas become inundated. For example, beachfront houses and condominiums may be lost because of shoreline movement. Additionally, storm hazards increase with sea level rise, resulting in increased damages and increased expenditures for repairing damages. Consequently, Set B may include the expenditure of more funds in response to storm damages than Set A. The difference in the values of Set B and Set A is the economic impact of sea level rise. It is important to note that because the economic activities in Community X include trade with other communities, the economic impact of sea level rise may be felt outside Community X, in places that are not physically threatened by rising sea level.

Community X may be better off if it is able to anticipate sea level rise and prepare for it. Anticipation would result in a third set of economic activities, Set C. For example, by anticipating sea level rise, people may decide not to build certain beachfront condominiums because of the anticipated rate of shoreline movement. Were it not for the anticipation of sea level rise, the condominiums would have been built and subsequently lost (or protected at great expense). If the structures are not built, the money that would have been used to build them would be used for something else. The value of anticipating sea level rise is the difference in the values of Set C and Set B.

This example provides several important insights. First, the economic impact of sea level rise is measured by comparing two quantities: the values of the two sets of economic activities defined above as Sets A and

B. The choice of those economic activities to be included in the analysis is important. If a particular economic activity is not affected by sea level rise, then the activity may be excluded from the analysis without biasing the results. However, excluding from consideration economic activities that are affected by rising sea level leads to a partial analysis, as discussed below.

Second, the economic impacts of sea level rise will depend on the actions people take in response to their changing environment. The actions people take will define, in part, how activity Set B differs from Set A. Consequently, the consideration of people's behavior is a critical aspect of this analysis.

Finally, this example highlights that the value of anticipating sea level rise is primarily a function of how anticipation changes people's behavior. If in the above example anticipation had no effect, then the resulting economic activities of Set C would be identical to Set B, and anticipating sea level rise would have no value. Therefore, to estimate the value of anticipating sea level rise, an assessment is required of what people might do (individually and collectively) if they knew that sea level was going to rise and had time to prepare for it.

Analysis of Economic Impacts

The objective of this economic analysis is to estimate the impact of sea level rise from the viewpoint of a community or study area. As described above, the study area carries on a set of economic activities over time that produce "net economic services." Net economic services (NES) can be thought of as the returns to a set of investments (gross services) minus the costs of the investments. Sea level rise may affect NES over time by altering the returns and costs of investments that are made in the study area and altering the mix of investments made in the study area. The second mechanism can be considered a feedback response whereby falling returns and increasing costs lead to reductions in future total investment. As explained below, property values may be used to measure NES.

This section is divided into three parts. First, the components of NES and the methods for measuring the components are presented. Then follows a discussion of the behavioral assumptions that drive the simulation of investment decisions over time. Finally, the section concludes with brief remarks on the economic impacts not captured by the analysis.

Components and Measurement of Net Economic Services. The development of the components of NES can be illustrated using an example of a house owned by an individual. The individual derives a certain level of satisfaction[2] from owning his house, which includes his valuation of the land, the capital (the structure), and all its amenities.[3] In any given year, call it year j, the individual derives some net economic services equal to NES_j. This quantity is equal to the gross services or returns derived (S_j) minus the costs of keeping the house (H_j). Therefore, NES_j for the individual is defined as follows:

$$NES_j \equiv S_j - H_j. \tag{7.1}$$

S_j equals the value the individual places on the use of his house, its location, neighborhood, and other amenities. These gross services can be likened to the amount the individual would be willing to pay in rent each year for the use of his house.

The costs of keeping the house include primarily maintenance and repair. For the purposes of this analysis, these costs have been broken down into three categories: costs of maintenance to cover routine depreciation; the costs of storm and flood damage; and the costs of actions taken to prevent, mitigate, or respond to the physical impacts of sea level rise (what are referred to in this volume as PMR activities).

Sea level rise may affect both gross services and costs. For example, a house may be located in a community near a beach. The current services derived from the house include the value of being close to the beach. With a rising sea level, the beach may be lost to erosion and rising water levels. As a result, the services derived from the house will fall by the value the user of the house placed on the beach.

The effect on costs can be seen more directly. Increasing storm surge elevations will cause increasing amounts of damage. Over time, the costs of repairing and maintaining the house will increase. If PMR actions are taken, the costs of these actions must be also considered. Finally, shoreline movement may affect both services and costs. If a house is lost to shoreline movement, both its future services and its future costs are eliminated.

To estimate the NES in a given year for a community, the NES from each of the individual properties can be added together. When summing across properties, double counting must be avoided. Using the beach as an example, its value is reflected in the services derived from the homes of individuals who use the beach.[4] It is not appropriate to estimate the service value of the beach separately and then add it to the service values of the homes. This would be double counting. The same is true for all other nonmarket amenities (such as parks). However, it is appropriate to add up the services derived from each of the individually owned properties.

When expanding the calculation of services from an individual to a community, a new term is added to the equation, namely, new investment. In the case of the individual, he only reinvests in his existing property, but in a community, new structures are required to serve the growing population. In the year in which new structures are built, the cost of construction is counted as a cost from the viewpoint of the study area. The new structures subsequently produce services during their lifetimes. Therefore, for a community, a term reflecting new investment is added (NI_j) and equation (7.1) is expanded as follows:

$$NES_j \equiv S_j - H_j - NI_j. \tag{7.2}$$

To aggregate the net services for a community over a period of time, the present value of the time stream of NES values for each year is estimated using a chosen discount rate. The discount rate reflects the relative value of dollars in different time periods; that is, a dollar next year is worth less than a dollar this year.[5] The choice of discount rate will, of course, influence the resulting estimate of the present value of NES over time.

Because the evaluation of NES for a community covers a finite period of time, a final term must be added to the calculation. This added term reflects the value of the capital stock at the end of the period, that is, those things with remaining useful lives. For example, a building may be built in the final year of the analysis. The cost of this new investment is counted in the estimate of NES for that year. However, the future services from the building are not counted because the analysis only examines a finite set of years. A quantity must be added that approximates the net value of the remaining life of the property; call it remaining capital stock (CS).

Equation (7.2) can now be expanded to include all the necessary terms, evaluated over time. Using the symbol PV(●) to indicate the present value of a finite stream of values over time, the expression for net economic services becomes:

$$PV(NES) = PV(S) - PV(H) - PV(NI) + PV(CS). \tag{7.3}$$

The first term to the right of the identity sign is the present value of gross services. The second and third values are the present values of the costs. The final term is the present value of the capital stock term. Next, the individual components of NES, starting with the identification of those items that contribute to NES, will be measured.

Within a study area, all articles of value can be thought of as producing a stream of services (e.g., a house produces housing services). To assess the impact of sea level rise, all those articles whose services or costs would be influenced should be included in the analysis. The exclusion of items whose services or costs are adversely affected will result in an underestimation of the impacts. The exclusion of items not affected by sea level rise does not result in bias.

The general list of inputs to the production of economic services includes land, capital, and labor. Both land and capital are important to include because they are fixed in location and directly affected by sea level rise. Shoreline movement can result in the loss of productive land and the capital improvements built on the land. Increased storm surge elevations will cause increased damages to structures during flooding, resulting in increased expenditures to maintain the building. These increased risks may reduce capital investment in the future (relative to levels that would have prevailed in the absence of sea level rise), resulting in a reduction in economic services.

Labor may also be affected, in terms of both supply and productivity. With increasing flood and erosion hazards in a coastal area, fewer individuals may choose to live and work there.[6] From the standpoint of the community, what is lost from a reduction in the use of labor is the value of the productive capacity of the labor minus the cost of the labor. Even if the amount of labor remains unchanged, its productivity may decrease. For example, more frequent interruptions due to flooding may reduce the average number of working days in a year, potentially affecting productivity.

Finally, important nonmarket amenities are likely to be affected, most notably beaches. If a beach is lost, the reduction in recreational opportunity is clearly a cost attributable to sea level rise.

To measure the net services produced by these various items, the analysis begins with observed market values of privately owned properties in the study area. Property values reflect the market's assessment of the present value of all future NES derived from a property. Included are people's valuations of nonmarket amenities such as beaches and parks. Additionally, for commercial properties, land values reflect the present value of future profit streams, including the appropriate estimate of the value of labor in excess of its costs.[7] Therefore, property values form a comprehensive measure of the market's expectations of future NES.

To estimate impacts, the current (and estimated future) property values were transformed into streams of gross services and costs. The impact of sea level rise on these streams was assessed directly.

The stream of gross services is affected by shoreline movement (which eliminates productive land and buildings) and by reductions in future economic activity. The cost stream is influenced primarily by changes in storm damage but also by the cost of community PMR actions and by reductions in future rates of new investment. The costs of routine maintenance were defined as the rate of depreciation of the structure times the value of the structure and are assumed to be constant.[8]

Sea level rise causes storm damage to increase because storm surge elevations will increase. The data required to calculate storm damages include: storm surge elevations and frequencies, the locations of high-wave-energy storm surge, topographical data, number of structures by location, the value of structures, and depth-damage functions (which relate the damage to a building to the depth of the flood above the first floor of the building). The storm surge and topographical data were obtained from the analyses of the direct physical effects of sea level rise reported earlier in this book.[9] Land use data were collected from a variety of sources for each study area.[10] Empirically derived depth-damage functions were used to calculate the value of storm damage to structures (including high-energy storm surge damage).[11]

The costs of storm and flood damages were calculated on an expected value basis. The total expected damage was computed by multiplying the damage from each storm type (e.g., a 100-year storm) times the probability of the storm occurring in any given year. The expected value of damages is analogous to an actuarily fair premium for insurance that would cover 100 percent of flood losses. This quantity reflects the true cost of the risk of storm damages on an annual basis. Of course, in any given year, a damaging storm may or may not occur. Consequently, the actual storm damages in any given year will rarely equal the expected value of storm damages. However, over a long period of time, the total damage experienced would approach the total expected value of damages, making the expected value an appropriate valuation of flood risk for the purposes of an economic analysis such as this.

It should be noted that an alternative approach to estimating storm damages is to simulate individual storm events over time. On average, the results of the simulation approach would be very similar to the expected value approach taken here. Nevertheless, the approaches would differ in an important way. Because severe storms (e.g., a 100-year storm) cause significant damage, the post-storm time period presents an opportunity to anticipate future sea level rise by significantly altering land use patterns. The expected value approach does not address this possibility and consequently results in an underestimation of the value of anticipating sea level rise. (The expected value approach was adopted here because of its relative simplicity from a computational point of view. The simulation of storm events was beyond the scope of the computing resources available for this effort.)

The costs of community PMR actions (e.g., seawalls and levees) were estimated from the unit costs provided by Sorensen et al. in Chapter 6. Insufficient data were available to simulate PMR costs on a per structure basis; consequently, individual PMR responses are omitted from this analysis.

The final component of NES is the amount of new investment occurring over time. New investment, by land use, is simulated to be driven by population changes within the study area. Detailed community development plans were used to project development to the year 2000. After that time, local, regional, and national population growth estimates were utilized. All land use was projected to increase at the rate of population growth after 2000, except for large special structures (such as the refining complex in Texas City, Texas), which were assumed to remain constant in size.

As described in the next section, the manner in which these components of NES change over time is driven by people's behavioral responses to sea level rise. In general, the shifts in land use and development are not devastating for the study area as a whole. For individual locations in the study area, however, simulated changes in land use in response to rising sea level can be quite significant.

Before turning to the discussion of the behavioral assumptions that drive the allocation of investment dollars and the choice of PMR actions over time, a technical consideration regarding the social value of capital investment must be mentioned. When individuals invest in a house or a commercial property, they evaluate the services derived from that property at their own private discount rate. It is often argued that the evaluation of economic activity from society's perspective should use a different (generally believed to be a lower) discount rate. Because of the divergence between social and private discount rates, the marginal value of an investment dollar is greater than one. Consequently, knowledge of the marginal value, or shadow price, of investment is required to estimate accurately the true NES over time from the social perspective of the study area. The calculation of this shadow price is particularly important because changes in investment are important responses to rising sea level.[12]

Behavioral Assumptions. A key component of this analysis is an assessment of how individuals, firms, and public bodies would respond over time to rising sea level. Models of rational economic behavior, as well as other models, have been applied to the question of how people respond to natural hazards such as floods and earthquakes.[13] The results of these investigations invariably demonstrate that people do not respond to risks from natural hazards in a manner consistent with models of rational behavior. Consequently, the assumption of rational behavior was rejected for the purposes of this analysis.

Once rationality is rejected as an adequate representation of human behavior, little is left in the way of quantitative bases for describing likely responses to the phenomenon of sea level rise. Nevertheless, a simple approach was developed by dividing behavioral responses into two types, which are simulated separately: the changes in investment decisions made by individuals and coordinated community PMR responses. The characterization of each type of behavioral response is discussed separately.

Individual investment decisions dictate the amounts of funds each year that will go toward reinvestment in existing properties to cover operation and maintenance costs, expenditures to fix storm damages, and investment in new development. For two reasons, the response of individuals is modeled as a slow, incremental process. First, the sea level rise phenomenon will unfold slowly. People will slowly adjust their behavior as their perceptions of the risks posed by the phenomenon develop. Barring major efforts on the part of government bodies (perhaps in concert with the scientific community) to influence people's actions (e.g., through land use regulation), it is likely that people will change their habits very slowly. Large, identifiable catastrophic events are not part of the unfolding sea level rise phenomenon; consequently, natural events will not jolt people's actions in a discontinuous fashion.

The second reason why an incremental approach is appropriate is that the impact of sea level rise, although important, is only one factor affecting the use of coastal areas. Coastal areas are used despite their hazards for a variety of economic and cultural reasons. Although the risk of storm damage may double or even quadruple with sea level rise, these costs remain only one factor affecting the use of the coastal environment. For example, in the Galveston case study, the annual cost of depreciation was estimated to be over 30 times the cost of expected annual storm damage (storm damage is low in part because of the extensive protective structures that have been built). Consequently, one would expect only small shifts in investment behavior as a consequence of the slowly increasing risk from storm damage. Of course, large increases in the rates of erosion and in annual risk of storm damage may have major consequences for portions of the study areas. As a whole, however, the general economic viability of the two coastal cities examined in this project is not threatened.

The small shifts in investment behavior were estimated by comparing the simulated condition of the study area over time to a reference case of economic development. The reference case was constructed under assumptions that there is no sea level rise and economic growth takes place as indicated by local community development plans and projected population growth. The reference case is characterized over time by the total market value of developed properties within the study area and the

Table 7-1. Summary of Simulated Private Investment Behavior

Quantity Estimated	Basis of Estimate
Step 1	
Initial estimate of total investment funds	Rate of investment in reference case
Adjustment for population growth	Census Bureau projections
Allocation of funds among	Allocation in reference case
Reinvestment	
Damage repair	
New investment	
Step 2	
Adjust total investment and allocation among investment types to reflect perceived increases in risks due to sea level rise	People's simulated perceived risks relative to reference case risks
Step 3	
Adjust damage repair investment to reflect actual damages	Damages simulated to be experienced

total amounts of funds expended on new investment, reinvestment (maintenance), and storm damage repair. As the actual case (e.g., the medium sea level rise scenario) begins to deviate from the reference case, people's investment behavior is simulated to shift away from the pattern characterized in the reference case.

Investment behavior was simulated in three steps, as summarized in Table 7-1. As the first step, the total amount of investment was calculated using the reference case as a guide. Total investment funds were initially set equal to the rate of investment per value of existing structures (determined from the reference case) times the value of existing structures in the actual case. This initial quantity of funds was then adjusted to reflect that new investment in structures is influenced heavily by population growth in the long run, which will deviate from the reference case only marginally (if at all) through changes in migration patterns in response to sea level rise. Consequently, a feedback was provided, whereby the rate of new investment is adjusted upward in proportion to the degree to which the existing structural values fall short of the values attained in the reference case. This feedback is important because it is a mechanism via which perturbations in the growth path of the community are dampened, allowing growth to approach the reference case values over time if the cause of the perturbation is eliminated. This adjusted level of investment funds was initially allocated among reinvestment, damage repair, and new investment in similar proportions to the reference case.

The second step was to compare people's perceived damages with the reference case damages. If the perceived rate of damages is equal to the rate in the reference case, then the initial allocation among investment types is used. However, with rising sea level, the perceived damages will generally exceed the reference case damages (as a percent of total property value) as people slowly perceive changes in risk.[14] Consequently, a greater proportion of the available investment funds is required to cover damages. These funds must either be taken out of new investment and reinvestment, or the total amount of investment must increase.

The propensity of individuals to increase total investment in response to increasing damages is unclear. A variety of assumptions were investigated, including a decrease in total investment funds equal to one-half the increase in storm damage risk, no changes in total investment funds, an increase in investment funds equal to one-half the increase in storm damage risk, and an increase in investment funds equal to the increase in storm damage risk. The first two approaches resulted in significant reductions in total property values relative to the reference case by 2025. These reductions appeared to be too large to be realistic, particularly in light of the fact that during this period, most of the sea level rise costs are small relative to other factors affecting investment. The last approach results in no change in total property values relative to the reference case; that is, people continue to build everything they would have built in the absence of sea level rise. The third approach results in plausible changes in investment behavior as a function of changes in perceptions of risk and was adopted for use in this analysis.

Clearly, a more sophisticated and empirically validated model of investment behavior would be preferable. If people's behavior in response to their changing perceptions would in fact look more like the last approach, then the estimates of the impacts of sea level rise reported here are biased downward. If investment behavior would look more like the first approach, then the estimates are biased upward. Biases in the estimates of the value of anticipating sea level rise as a consequence of this behavioral assumption move in the opposite direction.

The third and final step in simulating investment behavior was an adjustment reflecting the fact that people's perceptions of risk may be incorrect. As the result of the second step, investment goals have been set and funds have been committed to each of the three investment types. However, people underestimate damages because they underestimate the rate at which the sea level is rising and hence underestimate their risk. In order to meet the investment goal of covering a certain proportion of damages, the damage investment must be increased. Because damages occur probabilistically, people may not attribute these increased costs to sea level rise. Instead, increased damage expenditures (over expectations) may be attributed to unusually bad weather or other factors. Funds are not taken away from new investment or reinvestment because these funds are assumed to be committed. Instead, new funds are assumed to be added. By initially underestimating damages, the total investment increases, and the relative distribution of investment funds among the competing uses is altered.

A general model of urban development would be a useful extension of this method of simulating investment behavior. The current approach is clearly only a partial analysis because the wide range of alternative investment opportunities is not considered. However, this method results in shifts in investment behavior that move in the right direction at plausible rates. By rejecting the notion of optimal investment decisions in favor of incremental changes over time, the analysis provides an aspect of realism.

Community PMR actions were simulated separately from private investment decisions. Again, economic models of rational behavior were rejected as descriptors of likely responses. The actual experience of Galveston Island provides a good example of the misleading results that would be obtained by assuming rational economic behavior.

In the mid-1970s, the U.S. Army Corps of Engineers proposed the construction of a seawall to protect most of Galveston Island from bayside flooding.[15] Looking only at the costs of building the project and the benefits in terms of reduced storm damage to existing properties, the seawall was estimated to be beneficial and, based strictly on rational criteria incorporating quantifiable consequences, should have been built. However, the community rejected the proposal. The reasons for the rejection were not researched for this project but may include factors such as the inability to cover the community's share of the costs, environmental concerns, or possibly a belief that owners will not have to bear the full cost of damages because of the availability of disaster relief funds. In any case, the relationship between the cost of the protection project and the quantifiable benefits of the project in terms of reduced storm damage and erosion loss is insufficient for purposes of modeling the implementation of community PMR actions.

A more detailed model of community decision making could prove useful for this analysis. Such a model would describe communities' concerns and the decision process they go through when undertaking large protection projects. Numerous projects have been built by communities with the assistance and support of the Army Corps of Engineers. The data on the numerous projects built and the various projects rejected could be used to validate such a model.

Unfortunately, the resources and time available for this project did not allow a detailed examination of community response behavior. Instead, three basic types of PMR actions were defined, and the choice and timing of the actions were varied. The types of action are: stop or reduce the rate of shoreline movement through the use of revetments, levees, or other means; eliminate the threat of storm surge (up to a given elevation) through the use of seawalls and levees; and reduce or prohibit investment in given areas by "promulgating land use regulations." Other options not considered may include changes in building codes, beach nourishment, off-shore breakwaters, reclamation, and others.

The community PMR actions were assumed to be taken in various locations within the study areas at various times. Seawalls and high levees were used in threatened areas with high development density and high property values such as the Charleston Peninsula and Galveston Island. Revetments and low levees were used in places with medium development density that are threatened by rapid shoreline retreat. Land use regulations were applied in areas of significantly increasing hazard that were of low density.

The potential timing of the initiation of the PMR actions was divided into near term (1980-2010), medium term (2020-2050), and long term (2060-2080). In general, PMR actions were assumed to occur in the medium and high sea level rise scenarios, in the long and medium terms, respectively. These time frames are reasonable because at these times the physical changes caused by sea level rise would be clearly distinguishable from routine background variations.

By anticipating future sea level rise, the choice and timing of these PMR actions would be altered. This effect of anticipation is discussed in the next section. Under the assumption that no major intervention on the part of the federal government is undertaken to influence communities to make rapid responses to sea level rise, the community PMR actions simulated in this analysis are plausible representations of what may, in general, occur. The PMR actions are not chosen to be optimal (in the sense of maximizing net benefits) but instead are chosen to represent the major courses of action available to communities and the general time periods in which they are likely to be taken.

An example of a PMR action is shown in Table 7-2. To define the action, the action type must be identified as one of the three possible actions discussed above. The location in the study area that would be influenced by the action is also defined. Capital and operation and maintenance (O&M) costs are provided in terms of 1980 dollars. Finally, the applicable scenarios and times at which the action is taken are provided. The example in Table 7-2 is the modification of the Texas City levee system to protect areas of Texas City and La Marque from the increasing storm surge elevations found in the high sea level rise scenario. As shown in the table, the action is assumed to be taken in 2070 in both the medium and high sea level rise scenarios. As discussed below, to evaluate the value of

Table 7-2. Sample Data for Specifying a Community PMR Action

Community PMR Action Data	
PMR Action Type	Eliminate storm surge below a given elevation: 22 ft (6.7 m)
Location	Portion of Texas City/La Marque that becomes vulnerable to storm surge in high sea level rise scenario
Capital Cost	$100 million
O&M Cost	$0.1 million per year
Applicable Scenarios and Years Taken	Medium scenario; taken in 2070 High scenario; taken in 2070

anticipating sea level rise, the timing or the choice of the actions taken is altered.

Impacts Not Captured by This Method.   As described above, to the extent which this method excludes economic activities that are affected by rising sea level, the results will be of a partial nature. In three major areas, potentially important activities were omitted, resulting in an underestimate of the economic impacts of sea level rise. First, the costs of saltwater intrusion were not estimated. The annual value of the reduced availability of potable water should be added to the results derived here. Because the groundwater used in the two study areas analyzed here is not significantly affected by saltwater intrusion, this bias is not serious for the cases presented below.

The second impact omitted was the potential loss of economic value supplied by the near-shore zone. The loss of beach areas was not explicitly incorporated into the analysis. If a beach is lost, the value of the recreational opportunity it would have produced is lost. This cost could be very large, particularly if many beaches were affected simultaneously, reducing the availability of substitute recreation. Additionally, changes in the populations of aquatic species were not addressed. If the sea level rises rapidly, such as in the high scenario, various commercially important species may be unable to adapt to this changing environment, resulting in additional economic impacts.

Finally, the analysis does not address impacts outside the study area. Changes in investment behavior may have positive or negative secondary impacts elsewhere. From the viewpoint of this study, the changes in the two cities as a result of sea level rise are unlikely to have significant consequences for the rest of the nation as a whole. However, if sea level rise has a significant impact on investment behavior in all coastal communities simultaneously, then the secondary effects could be considerable. This question warrants further consideration.[16]

Analysis of the Value of Anticipating Sea Level Rise

The most general way to think about the value of anticipating sea level rise is to ask what would happen if sea level rise were not anticipated and what would happen if it were. The answers to these questions are, by the nature of the analysis, uncertain. It is not known what will happen because it is not known how fast the sea level is going to rise. It is clear, however, that the value of anticipating sea level rise will depend, in part, on how rapidly the sea level actually rises. For example, if the high scenario is true, it is more valuable to plan for it ahead of time than if the low scenario is true.

Because the uncertainty about sea level rise is large, the uncertainty over the value of anticipating it is also large. To address this uncertainty, the approach used here was designed to produce separate estimates for each scenario of rising sea level. The results of the analysis should be interpreted as contingent estimates, such as: if the low scenario is true, then the value of anticipating sea level rise is $X million. Individual readers may decide for themselves the likelihoods of the various scenarios.

To estimate the value of anticipating sea level rise, the economic impact analysis method was augmented to incorporate alternative investment behaviors and community PMR actions. The economic impact analysis implicitly assumes that people act as though they currently believe that the sea level is rising at a rate less than or equal to historical trends. These beliefs are simulated to slowly approach the accurate perception of the sea level rise scenario being analyzed. If sea level rise is anticipated, these simulated investment and PMR behaviors will change. To estimate the values of these behavioral changes, the results of the economic impact analysis were used as a baseline of comparison. The reduction in the impact of sea level rise due to the simulated changes in behavior is the value of anticipating sea level rise.

Exactly how individuals and communities would behave in anticipation of sea level rise is uncertain. How people will behave will depend, in part, on how accurately future rates of sea level rise can be predicted and the level of confidence associated with the predictions. The role of federal, state, and local governments will be important, particularly in regard to their regulatory activities and economic incentives.

One method of modeling this behavior would be to develop optimal strategies to undertake in anticipation of sea level rise and to determine how people and communities should behave to minimize the adverse impacts of sea level rise. Such analyses are needed to inform the sea level rise debate. Of course, the definition of the optimal strategy (or strategies) is always constrained by the ability to quantify and value all the important impacts.[17] Consequently, the results of such analyses of optimal solutions should be viewed as guides to decisions, not as definite answers.

Rather than ask how people should act, this analysis examined how people probably would act. The differences in approach and results are considerable. Optimal behavior was not estimated; instead (for the same reasons discussed in the previous section), individual behavior was assumed to change slowly in an incremental manner. Additionally, individuals were assumed to be rather near-sighted in their investment decisions. Whereas an optimal preparation strategy would utilize all available information about future sea level rise, individual investment decisions are assumed to have horizons of only 10 years. This limited decision horizon for individuals was adopted as representative of a variety of factors, most principally, the potentially high discounting of the future by individuals, the uncertainty associated with long-range predictions, the inability or unwillingness of individuals to incorporate uncertain information into their decisions, and the costs associated with obtaining or developing information. Consequently, individuals were simulated to improve their investment behavior by preparing only for the true increases in risks during the coming decade.

Investment behavior is improved by reducing investment funds (relative to the base case) in areas of increasing risks. The investment fund allocation procedure described in the previous section was utilized. However, whereas in the base case individuals systematically underestimate risks, by anticipating sea level rise they accurately assess risks in the coming decade.

Community PMR actions also change because of the anticipation of sea level rise. Primarily, PMR actions were assumed to be taken in anticipation of increasing hazards instead of in response to increasing hazards. Consequently, the timing of the building of protective structures and the promulgation of zoning restrictions was assumed to be 20-40 years earlier than assumed in the economic impact analysis.

The primary weakness of the approach taken here is its subjective nature. A variety of assumptions are made about how individuals and communities might prepare for future sea level rise. Although these assumptions are both internally consistent and plausible, they have not been empirically validated; consequently, they should not be ascribed predictive ability. Instead, the results presented here provide an estimate of the order of magnitude of savings that could be realized by anticipating sea level rise.

In many respects, our methods biased the estimates of the value of anticipating sea level rise downward. No advancement in the state-of-the-art of protective structures was assumed. Opportunities to take advantage of rises in sea level were not examined (e.g., in siting port facilities). Finally, the actions simulated to be taken in anticipation of sea level rise are only small shifts away from the behavior simulated in the economic impact analysis. In fact, preparatory actions could be much more comprehensive, particularly if the federal government were to take a strong leadership role by restructuring incentives for development in the coastal zone to be more appropriate for a rising sea level.

RESULTS

Results for the Charleston and Galveston case studies are reported separately below. A variety of values was used for the private and social discount rates and other parameters. The results presented below represent one set of assumptions and parameters, including a private real discount rate of 10 percent per year; a real appreciation rate of property of 2 percent per year; total storm damages equal to twice the damages to privately owned structures to account for damage to contents, publicly owned structures, and economic disruption;[18] and real social discount rates of 3 percent, 6 percent, and 10 percent per year.

Charleston, South Carolina

The Charleston study area, shown in Figure 7-1, includes the city of Charleston and portions of North Charleston, Mount Pleasant, Sullivans Island, James Island, and Daniel Island. The developed portion of the study area (the entire area excluding Daniel Island) was divided into 37 subareas of approximately 2.2 sq km (0.85 sq mi) each. The subareas, identified from Charleston County property tax assessment maps,[19] represent fairly homogeneous areas of land use.

The following information was obtained for each subarea: the physical impacts of each sea level rise scenario in terms of storm surge (elevation and frequency) and area loss due to shoreline movement,[20] topography,[21] the number and average value of existing structures by land use type,[22] and anticipated land use changes by the year 2000.[23]

The results for the study area are reported in terms of four quantities: the market value of structures over time, the expected value of storm damages over time, losses due to shoreline movement by decade, and the present value of net economic services in 1980 dollars. The changes in market values represent, in part, changes in investment in response to sea level rise. Also, changes in the expectations of storm damage are important. Potential losses due to shoreline movement play a more important role in the Charleston study area than in the Galveston area, which is mostly protected by existing seawalls and levees. Finally, the aggregate economic impacts are summed up in the NES estimates.

Figure 7-2 displays how market values are affected by rising sea level. Curve *A* displays the trend case and shows steadily increasing market values over the next century from the current $1.27 billion. The high scenario without anticipation (Curve *B*) diverges from the trend case beginning in the year 2000. Beginning in 2020, the community is simulated to take actions to reduce the losses to shoreline movement by protecting the Charleston Peninsula and areas west of the Ashly River. However, without anticipation of sea level rise, these measures are assumed to be less than totally effective because the rate of sea level rise is underestimated. Consequently, additional actions that could eliminate the shoreline movement problems in most areas are simulated to be taken

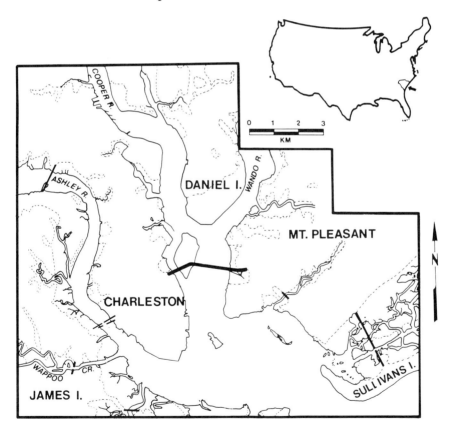

Figure 7-1. Map of the Charleston case study area. *(Map supplied by Research Planning Institute. Columbia, S.C.)*

by 2060. Between 1980 and 2060, some areas will have been developed that cannot be protected, resulting in additional losses after 2060.

Curve C shows the results of the high scenario with anticipation of sea level rise. In anticipating sea level rise, several coordinated responses are simulated to be taken. First, the Charleston Peninsula is protected from shoreline movement by a levee or seawall constructed in 2010. This structure is assumed to be effective in stopping shoreline movement throughout the highly developed peninsula and in providing some protection from storm surge. Second, the area west of the Ashley River is divided into two parts. The area near the Wappo Creek is protected from shoreline movement with a low levee system, and new development is intensified. The area north of there, near the first bend in the Ashley River, is presumed to go unprotected, and new development is assumed to be prohibited. Third, Mount Pleasant is presumed to be developed in a manner that minimizes subsequent loss to shoreline movement, possibly with the use of revetments. Last, investment in Sullivans Island, an area

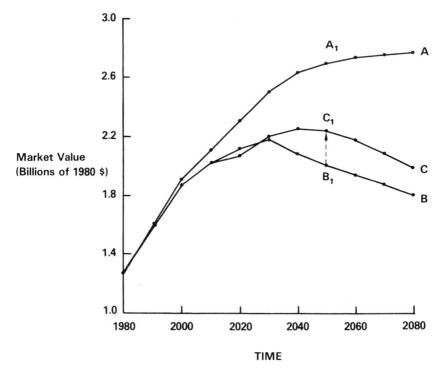

Figure 7-2. Market values in the Charleston case study area over time for three scenarios. (A) trend rate of sea level rise; (B) high sea level rise scenario without anticipation; and (C) high sea level rise scenario with anticipation.

of rapidly increasing storm hazard and shoreline movement, is presumed to be reduced significantly.

Even with these various actions, however, market values continue to decline. This continuing decline is due in part to the simulated choice to reduce investment in certain areas but is also due to the increasing risk of storm damage and continued shoreline movement.

The significance of Figure 7-2 is that by anticipating sea level rise, development and protective measures can be undertaken to reduce the losses from sea level rise. For example, by 2050 in the high scenario, anticipating sea level rise results in over $200 million in additional market value relative to the case when sea level rise is not anticipated (point B_1 to point C_1 in Figure 7-2). However, even with anticipation of sea level rise, point A_1 (projected market value in the trend case) is not attained.

The community PMR responses in Charleston would require significant cooperation among a variety of jurisdictions. Protecting the peninsula would involve both the cities of Charleston and North Charleston. Additionally, there is a federally owned naval facility on the northeastern portion of the peninsula. The area west of the Ashley River includes James Island, the city of Charleston, and some unincorporated land controlled by the county. Also, a new highway, the Mark Clark Expressway, is anticipated to be constructed west of the Ashley River. Therefore, an effective community response in the Charleston study area would require a mechanism for performing regional planning. The time lags involved in establishing and developing such a regional planning authority could be an important factor affecting the magnitude of the impact of sea level rise.

Figure 7-3 shows the pattern of storm damage for three scenarios. The trend scenario (curve A) displays a slowly increasing amount of storm damage over the next century, driven primarily by increasing market values. Curves B and C, showing the high scenario, diverge significantly from the trend scenario by the year 2000. Without anticipating sea level rise (curve B), the storm damage continues to rise through 2030. By that time, shoreline movement is causing such large losses that the total storm damage actually begins to decline because fewer structures remain to be damaged. Although the risk of storm damages is increasing, the total value of structures at risk is decreasing, resulting in lower aggregate damages. Damages continue to decline through 2060 as the area continues to experience large losses due to shoreline movement. By 2060, the shoreline movement problems are assumed to be arrested and total storm damages begin increasing.

The high scenario with anticipation (curve C) diverges from curve B in 2010 when coordinated protective actions are simulated to be taken.

Storm damages decline through 2020 but increase substantially thereafter. By 2080, the storm damage in the with-anticipation case exceeds the without-anticipation case by a considerable amount. This counterintuitive result is caused by the protection of areas from shoreline movement. Areas west of the Ashley River are simulated to be protected by a low levee that does not significantly reduce storm surge. As a consequence, storm damages increase significantly with sea level rise. From the standpoint of storm damages, this response is probably not the best possible response. However, given the medium density development projected for the area, it is a likely response.

Figure 7-4 shows the considerable reductions in shoreline movement losses simulated to be realized with anticipation in the high scenario. Without anticipating sea level rise, shoreline movement losses grow rapidly through 2040. By 2060 actions are simulated to be taken to reduce losses. Even so, shoreline movement losses average $57 million per decade from 1980 to 2080. Anticipating sea level rise reduces shoreline movement losses by an average of approximately $5 million per decade over the same period.

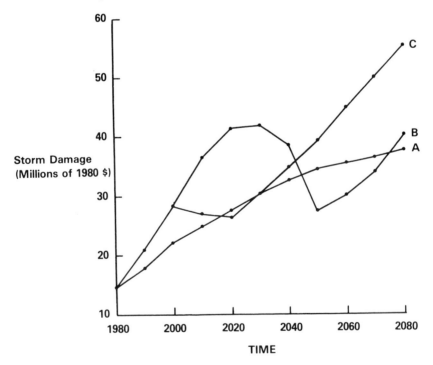

Figure 7-3. Storm damage in the Charleston case study area over time for three scenarios. (A) trend rate of sea level rise; (B) high sea level rise scenario without anticipation; and (C) high sea level rise scenario with anticipation.

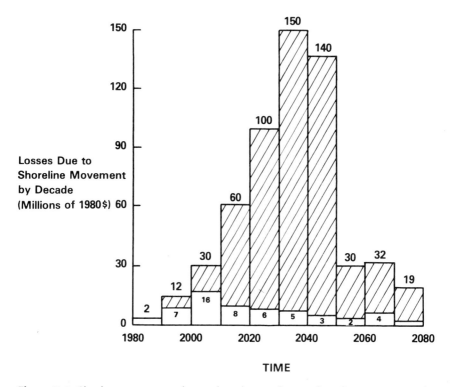

Figure 7-4. Charleston case study results—losses due to shoreline movement by decade. High scenario without anticipation = upper number. High scenario with anticipation = lower number. Savings attributable to the anticipation of sea level rise = shaded area.

The diverse impacts of changing market values, land use, shoreline movement, and damages are summarized in the NES calculation. Table 7-3 reports the NES values for the trend scenario and the high scenario, with and without anticipation of sea level rise for the period 1980-2025. The economic impact of the high scenario is estimated by subtracting the NES for the high scenario (without anticipation) from the trend scenario NES. From Table 7-3 it is seen that the economic impact of the high scenario evaluated at a real 3 percent discount rate is $1,065 million. Anticipating future sea level rise could reduce this impact by over 60 percent, and the value of anticipation is estimated at $645 million. The values are somewhat smaller when a 6 percent or 10 percent discount rate is used.

Tables 7-4 and 7-5 summarize the estimates of economic impacts and value of anticipating sea level rise for each of the scenarios. The economic impacts are much larger for the 1980-2075 period than for just 1980-2025, as would be expected. Because these values are present values

Table 7-3. Estimates of Net Economic Services for the Charleston Case Study Area, 1980–2025 (in millions of 1980 dollars)

	Real Discount Rate (in Percent)		
	3	6	10
Scenario			
A. Trend Scenario	5,395	2,840	1,730
B. High Scenario without Anticipation	4,330	2,570	1,665
C. High Scenario with Anticipation	4,975	2,685	1,675
Results			
Economic Impact (A-B)	1,065	270	65
Value of Anticipation (C-B)	645	115	10

Table 7-4. Economic Impacts of Three Sea Level Rise Scenarios in the Charleston Case Study Area at Three Discount Rates and for Two Periods of Time (in millions of 1980 dollars)

	Real Discount Rate (in Percent)[a]					
	1980–2025			1980–2075		
Scenario	3	6	10	3	6	10
Low	280	70	15	1,250	110	20
	(4.9)	(2.4)	(0.8)	(17.3)	(3.6)	(1.1)
Medium	685	165	40	1,910	305	50
	(12.0)	(5.6)	(2.2)	(26.5)	(10.1)	(2.8)
High	1,065	270	65	2,510	440	80
	(18.7)	(9.1)	(3.7)	(34.8)	(14.6)	(4.5)

[a]Values in parentheses report percentage of total net economic services estimated in the trend case.

Table 7-5. The Value of Anticipating Future Sea Level Rise for the Charleston Case Study Area, Contingent on Each of the Three Sea Level Rise Scenarios (in millions of 1980 dollars)

	Percentage Real Discount Rate (in Percent)[a]					
	1980–2025			1980–2075		
Scenario	3	6	10	3	6	10
Low	120	25	5	810	55	5
	(43)	(36)	(33)	(65)	(50)	(25)
Medium	340	50	10	1,180	160	10
	(50)	(30)	(25)	(62)	(53)	(20)
High	645	115	10	1,400	230	25
	(60)	(43)	(5)	(56)	(52)	(31)

[a]Values in parentheses report percentage of total ecomonic impact.

Table 7-6. Economic Impact and Value of Anticipating Sea Level Rise for the High Scenario, 1980–2075: Charleston Case Study Area

Portion of Charleston Study Area	Economic Impact of High Scenario (Percent of Study Area)	Value of Anticipating High Scenario (Percent of Study Area)
Peninsula: Charleston and North Charleston	900 (36%)	950 (68%)
West Ashley/James Island	685 (27%)	310 (22%)
Mount Pleasant	600 (24%)	80 (5.7%)
Sullivans Island	325 (13%)	60 (4.3%)
Total Study Area	2,510 (100%)	1,400 (100%)

Note: Values are present values in millions of 1980 dollars evaluated at a real discount rate of 3 percent per year.

of streams over long periods, the discount rate has a significant influence on the outcome.

Table 7-4 indicates that even the low scenario will have impacts in excess of $1 billion by 2075 (evaluated at a 3 percent discount rate). Relative to the trend scenario, this is over a 17 percent reduction in the value of the economic activity in the study area over the 100-year period analyzed. (Percentage reductions are reported in parentheses in Table 7-4.) Table 7-5 reports that by anticipating sea level rise, the impacts can be reduced significantly. For example, by anticipating the low scenario, the study area could be $810 million better off, offsetting 65 percent of the economic impact of the low scenario.

Table 7-6 presents a breakdown of the results by four subareas. The peninsula area has the highest impact in the high scenario, $900 million. By anticipating sea level rise, the actions simulated to be taken more than offset this adverse impact. The peninsula is simulated to be protected by seawalls and low levees in the high scenario. This raises the question whether the peninsula would be better off with these protective measures, even without sea level rise. The results of our analysis indicate that the benefits of such protection would currently outweigh its costs by an order of tens of millions of dollars. However, the analysis presented here does not consider reduced access to the waterfront, a reduction in the scenic beauty of the area, or environmental impacts. Nevertheless, if faced with the high sea level rise scenario, major protective structures would be required to prevent the loss of large areas of the highly developed center of Charleston.

The West Ashley/James Island area has the second-highest impacts, $685 million. Without anticipation of sea level rise, significant new development would take place over the next 20-40 years that would subsequently be either lost to shoreline movement or subject to a greatly increased risk of storm damage. By anticipating sea level rise, development can be limited to those areas that can be easily protected with low levees. This strategy offsets nearly 50 percent of the impacts.

The Peninsula and West Ashley/James Island areas account for 63 percent of the impacts and 90 percent of the value of anticipating the high sea level rise scenario in the Charleston case study area. Although Mount Pleasant and Sullivans Island both suffer significant impacts, responses resulting in significant savings were not identified. However, for any regional preparation for sea level rise to be implemented, these areas would have to be involved because of the integrated nature of the transportation system and commerce in the area.

Galveston, Texas

The Galveston study area, shown in Figure 7-5, includes a portion of Galveston Island, Texas City, La Marque, San Leon, and some unincorporated areas. The study area was divided into 97 subareas of approximately 2.8 sq km (1.08 sq mi) each. These subareas are the same units used by Leatherman[24] to characterize the physical impacts of sea level rise. The following information was obtained for each subarea: the physical impacts of each sea level rise scenario in terms of storm surge (elevation and frequency) and area loss due to shoreline movement,[25] topography,[26] population in 1980 and projected population in 2000,[27] the number and average value of existing structures by land use type, and anticipated land use changes by the year 2000.[28]

The results for the study area are reported in terms of three quantities: the market value of structures over time, the expected value of storm damages over time, and the present value of net economic services. The changes in market values of land and structures indicate the extent to which investment in the area is reduced in response to sea level rise. Expected storm damages are reported because they are the major physical impact on the study area. The cost of storm damages is approximately one magnitude larger than the value of the land lost to shoreline movement. Finally, the estimates of net economic services (NES) sum up the total impact on the study area. NES estimates are provided for two time periods, 1980-2025 and 1980-2075. As indicated below, the NES values are sensitive to the choice of discount rate.

Figure 7-6 displays estimates of the market value of land and structures for three scenarios. The top curve (Curve A) represents the simulated results if the sea level rises at a rate equal to recent historical trends. Driven by population growth, the value of structures (currently $3.3 billion) is anticipated to grow steadily over the next century. Curve *B* represents the high scenario, analyzed without anticipation of sea level rise. The impact of sea level rise on market values is moderate from 1980

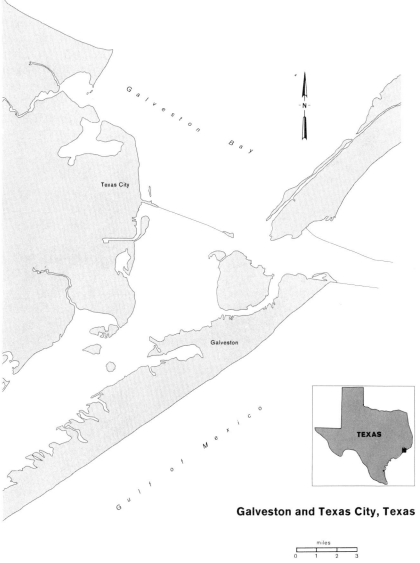

Galveston and Texas City, Texas

miles
0   1   2   3

Figure 7-5. Map of the Galveston case study area. *(Map supplied by Stephen P. Leatherman, University of Maryland.)*

to approximately 2030, after which time impacts are significant. The time period between 2020 and 2030 is a turning point for this scenario because at this time, the protected areas within the Texas City levee system and behind the Galveston seawall are simulated to become vulnerable to storm surge. However, without better information, it is assumed that the communities would not recognize this threat and consequently would not respond until later. Market values begin to fall because of the rising expense from storm damage. As events unfold between 2030 and 2050, people recognize their increased risk and by 2070, the levee system and seawall are simulated to be upgraded to provide the necessary protection. By that time, however, the increased risk of storm damage has had a significant impact. The rate of decline in values slows after 2070.

Curve C in Figure 7-6 depicts the high sea level rise scenario with anticipation. In the period 1980-2040, the market value is slightly below the without-anticipation case because the community is assumed to restrict investment in areas of rapidly increasing hazard. In the 2040 to

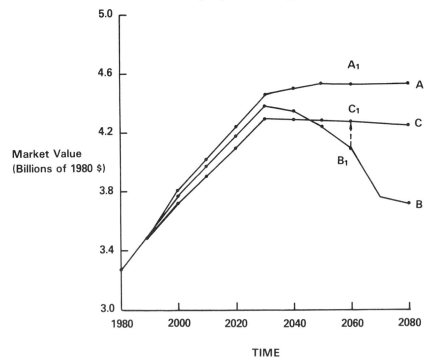

Figure 7-6. Market values in the Galveston case study area over time for three scenarios. (A) trend rate of sea level rise; (B) high sea level rise scenario without anticipation; and (C) high sea level rise scenario with anticipation.

2080 period, the market value exceeds the without-anticipation case (curve B) because the community is assumed to upgrade the levee system and seawall in anticipation of the increased hazard. For example, by 2060 in the high scenario, anticipating sea level rise results in approximately $200 million less reduction in market values of property (point C_1 versus point B_1). However, even with anticipation of sea level rise, point A_1 (market value with the trend sea level rise scenario) is not attained.

Figure 7-7 presents the storm damages over time for the same three scenarios. The trend case shows slowly increasing damages, in part because of the increasing market value. The high scenario without anticipation (curve B) displays how the risk of storm damage jumps up after 2020 and again after 2050. These jumps occur in part because of the discontinuous nature of the protective structures and in part because of the discrete function used to estimate storm damages.[29] In the 2020-2030 period, damages jump to nearly $40 million per year because the protected area becomes vulnerable to the 100-year storm.[30] By 2060 many of the protected areas would also become vulnerable to the 50-year storm, as is indicated by the jump in damages during that decade. By 2070 it is

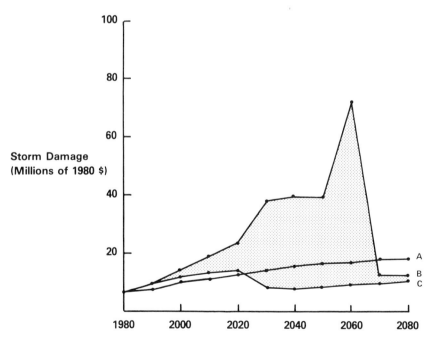

Figure 7-7. Storm damage in the Galveston case study area over time for three scenarios. (*A*) trend rate of sea level rise; (*B*) high sea level rise scenario without anticipation; and (*C*) high sea level rise scenario with anticipation.

Table 7-7. Estimates of Net Economic Services for the Galveston Case Study Area, 1980–2025

	Real Discount Rate (in Percent)		
	3	6	10
Scenario			
A. Trend Scenario	10,010	6,015	3,985
B. High Scenario without Anticipation	9,650	5,915	3,960
C. High Scenario with Anticipation	8,870	5,970	3,975
Results			
Economic Impact (A–B)	360	100	25
Value of Anticipating Sea Level Rise (C–B)	220	55	15

assumed that the community upgrades its protection systems and that risks decrease substantially.

With anticipation of sea level rise (curve C), the damages are initially below curve B because investment is reduced in areas of increasing hazard. Curves B and C diverge dramatically after 2020 when the upgrading of the protective systems is assumed to take place. The shaded area in Figure 7-7 is the total reduction in damages attributable to the anticipation of sea level rise.

The diverse impacts of changing market values, land use, and damages are summarized in the NES calculation. Table 7-7 reports the NES values for the trend scenario, and for the high scenario with and without anticipation, for the period 1980-2025. The economic impact of the high scenario is estimated by subtracting the NES for the high scenario without anticipation ($9.65 billion) from the trend scenario NES ($10.1 billion). From Table 7-7 it can be seen that the economic impact of the high scenario evaluated at a real 3 percent discount rate is $360 million. Anticipating sea level rise could reduce this impact by over 60 percent; the value of anticipation is estimated at $220 million. The estimates are much larger for the 2025-2075 period than for 1980-2025.

Tables 7-8 and 7-9 summarize the estimates of economic impact and value of anticipation for each of the scenarios. Because these estimates are present values of streams over long periods, the discount rate has a significant influence on the outcome. The values are substantially smaller when a 6 percent or 10 percent discount rate is used.

Table 7-8 shows that the economic impact of sea level rise in the Galveston study area may range from $555 million for the low scenario to nearly $1.9 billion for the high scenario through 2075. These impacts represent a reduction of 4.9 to 16.0 percent of the total present value of economic activity in the study area during that period. The impacts would

Table 7-8.   Economic Impact of Three Sea Level Rise Scenarios in the Galveston Case Study Area (in millions of 1980 dollars)

| Scenario | Real Discount Rate (in Percent)[a] | | | | | |
| | 1980-2025 | | | 1980-2075 | | |
	3	6	10	3	6	10
Low	115	30	10	555	70	10
	(1.1)	(0.5)	(0.2)	(4.9)	(1.2)	(0.2)
Medium	260	65	15	965	150	20
	(2.6)	(1.1)	(0.3)	(8.4)	(2.5)	(0.5)
High	360	100	25	1,840	220	35
	(3.6)	(1.7)	(0.6)	(16.0)	(3.6)	(0.9)

[a]Values in parentheses report percentage of total net economic services estimated in the trend case.

Table 7-9.   The Value of Anticipating Future Sea Level Rise for the Galveston Case Study Area, Contingent on Each of the Three Sea Level Rise Scenarios (in millions of 1980 dollars)

| Scenario | Real Discount Rate (in Percent)[a] | | | | | |
| | 1980-2025 | | | 1980-2075 | | |
	3	6	10	3	6	10
Low	25	6	1	245	27	2
	(22)	(20)	(10)	(44)	(39)	(20)
Medium	150	30	5	550	60	5
	(58)	(46)	(33)	(57)	(40)	(25)
High	220	55	15	1,110	130	15
	(61)	(55)	(60)	(60)	(59)	(43)

[a]Values in parentheses report percentage of total economic impact.

have been much larger were it not for the existing seawall and levee systems in the area. Leatherman assumed that the seawall and levee systems would be upgraded as necessary so that they would remain effective with a sea level rise. Consequently, in this analysis, they provide protection from the increasing frequency of storm surge at elevations below the minimum height of the structures. Only storm surges in excess of the minimum height of the structures were assumed to cause damage. If the structures were assumed to fail, then the impacts would be larger because storm damages would be larger. (The existing protective structures were also assumed to halt shoreline movement at their current locations.)

Table 7-9 shows that by anticipating sea level rise, its adverse impacts can be greatly reduced: impacts from the high scenario can be reduced by approximately 60 percent. Table 7-10 displays how the economic impact and value of anticipating sea level rise are distributed throughout the

Table 7-10. Economic Impact and Value of Anticipating Sea Level Rise for the High Scenario, 1980–2075: Galveston Case Study Area

Portion of Galveston Study Area	Economic Impact of High Scenario (Percent of Study Area)	Value of Anticipating High Scenario (Percent of Study Area)
Galveston Island/Bolivar Peninsula	620 (34%)	565 (51%)
Texas City/La Marque	950 (51%)	535 (48%)
San Leon	270 (15%)	10 (1%)
Total Study Area	1,840 (100%)	1,110 (100%)

Note: Values are present values in millions of 1980 dollars evaluated at a real discount rate of 3 percent per year.

study area for the high scenario. Both the Galveston Island and Texas City/La Marque areas are significantly affected by the high sea level rise scenario. By anticipating sea level rise, the impacts in these areas can be reduced substantially. In the Galveston Island area, the anticipation of sea level rise is simulated to result in earlier implementation of measures to protect the island from bayside flooding. By taking these actions before the risk increases substantially, damages are prevented.

The impact of the high scenario in the Texas City/La Marque areas is caused primarily by currently protected areas becoming vulnerable to storm surge. To protect the areas, the levee system is simulated to be extended, thus reducing the risk of storm damage. The anticipation of sea level rise is assumed to result in an earlier extension of the levee system than would otherwise occur. Additionally, there are unincorporated areas south of La Marque that are very low-lying and are not protected by the levee system. Under the high scenario, these locations would be inundated. These areas are expected to be developed over the next 40 years. By anticipating the high scenario, it is assumed that additional development after 1990 would not be undertaken in these very vulnerable areas.

The San Leon area is also very vulnerable to inundation under the high scenario, as is indicated in Table 7-10 by impacts of $270 million. Although this is only 15 percent of the impact for the entire study area, it represents 70 percent of the value of the economic activity simulated to be undertaken in the San Leon area. Anticipating sea level rise is estimated to have only a small benefit for San Leon because it was assumed that the medium to low density of development there was insufficient to justify large protective structures. Consequently, development patterns are simulated to shift away from the areas of increasing hazard, resulting in a slight reduction in impacts. If San Leon could be protected (possibly as a part of a larger protection effort for all of the Galveston Bay area), then the benefit of anticipating sea level rise would probably be higher.

SUMMARY

This chapter has presented the implementation of a method for assessing the economic impacts of sea level rise and has examined the value of anticipating sea level rise. Although significant work remains to be done, this work provides a step toward developing improved assessments of the economic and societal impacts of sea level rise.

As one would expect, coastal cities currently without protection are more vulnerable to sea level rise and have the most to gain by anticipating it. As shown in Table 7-11, the Charleston study area appears to be somewhat harder hit by sea level rise, primarily because of its current unprotected state relative to Galveston. Evaluated at a 3 percent real discount rate, the impact of the high scenario is 16 percent of the total NES in the Galveston study area for the 1980-2075 period. In Charleston, the impact is over 34 percent.

The estimates of the value of anticipating sea level rise must be considered to be somewhat speculative but nevertheless highly suggestive. The small, marginal shifts in behavior modeled here are found to have considerable value. As described in the text, the values reported here may be biased downward by the conservative assumptions employed.

Based on the results presented here, it appears to be justified to ask what should be done in anticipation of sea level rise. The analysis illustrates that the stakes are large and that appropriate preparation can significantly reduce adverse impacts. Consequently, investigating what we ought to do to prepare ourselves is clearly warranted. Choosing a

Table 7-11. Summary of Economic Impacts and Value of Anticipating Sea Level Rise in the Charleston and Galveston Case Study Areas

	Charleston Study Area		Galveston Study Area	
Scenario	Economic Impact	Value of Anticipation	Economic Impact	Value of Anticipation
Low	1,250	810	555	245
Medium	1,910	1,180	965	550
High	2,510	1,400	1,840	1,110

Note: Values are present values in millions of 1980 dollars evaluated at a real discount rate of 3 percent per year.

preparation strategy requires an analysis very similar to that described here. In addition, the uncertainty surrounding how much sea level will actually rise must be incorporated into the analysis.

Even with our current uncertainty regarding future sea level rise, the large potential impacts combined with the possible savings from preparing for sea level rise suggests that taking actions today should be considered. If actions are taken today (e.g., accelerating research or incorporating the possible future need for protective structures into current designs), they may turn out to be unnecessary if the sea level does not rise. Alternatively, if the sea level is rising as fast as indicated by the scenarios used here, we will be much better prepared 20 years from now and as a result will be much better off. Analysis and political debate are required to balance the risk of taking unnecessary actions against regretting 20 years from now that opportunities were missed. The estimates reported here provide a basis upon which the analyses and debate can be built and suggest that a resolution of this issue is very important to Charleston and Galveston.

NOTES

1. To the extent that investors and developers have knowledge of erosion and storm hazards and incorporate this knowledge in their investment decisions, development patterns will reflect people's perception and valuation of hazards. For examples, see J. R. Barnard, 1977, "Economic Costs Associated with Increased Flood Hazards from Urban Growth," in *Proceedings of International Symposium on Urban Hydrology, Hydraulics, and Sediment Control.* Bulletin 114. Lexington, Ky.: Kentucky University Office of Research and Engineering Services.
2. The specification of utility functions as measures of satisfaction were omitted from this analysis. Instead, dollars are used.
3. Amenities may include nice weather, scenic views, air quality, and others. For more detail on amenities see Douglas B. Diamond, Jr., and George S. Tolley, eds., 1982, *The Economics of Urban Amenities*, New York: Academic Press.
4. If people who use the beach live outside the study area, then the value of the beach will be underestimated because their property values reflecting their valuation of the beach will be excluded. This bias can be eliminated by incorporating the value that people living outside the study area place on the beach (estimated by other means).
5. The reader interested in discounting and discount rates may refer to E. J. Mishan, 1976, *Cost-Benefit Analysis*, New York: Praeger, pp. 199–218.
6. From the standpoint of the individual, sea level rise may reduce the attractiveness of a coastal community because of increased storm hazards. Consequently, the individual may decide to leave an area (or not come in the first place), whereas he or she would otherwise have lived there. In this case, the individual is worse off. This impact is not captured by this analysis.

7. Surplus value of labor refers to the "producers' surplus" or portion of the worker's value captured by the employer. For more on producers' surplus see Mishan, *Cost-Benefit Analysis*, pp. 55-64 (note 5).
8. The following rates of depreciation were used:

    Single-family houses: 2 percent per year, based on the assumption of a 50 year lifetime for homes generally used in property value assessments. See George F. Bloom, and Henry S. Harrison, 1978, *Appraising the Single Family Residence*, Chicago: American Institute of Real Estate Appraisers of the National Association of Realtors.

    Manufactured housing: 5 percent per year, based on an expected life of 20 years.

    Commercial/industrial properties: 9 percent per year. See William Williams, 1961, *The Measurement of the Impact of State and Local Taxation on Industrial Locations*, Boulder: University of Colorado, Department of Economics.

    Multi-family housing: 5 percent per year, assumed to be approximately divided between single family housing and commercial/industrial properties.

9. See Chapters 4 and 5.
10. Land use data were collected for each of four land uses: single-family houses, multi-family houses, manufactured housing (e.g., mobile homes), and commercial/industrial properties.
11. See Don Friedman, 1975, *Computer Simulation in Natural Hazard Assessment*, Institute of Behavioral Science, NTIS PB-261-755, Boulder: University of Colorado.
12. Based on the principles discussed in Lind, the marginal value of investment is estimated by putting all the returns from an investment in consumption units. Assuming that sea level rise has little or no impact on the savings rate nationally, the marginal value of investment can be expressed in terms of the private discount rate (r), the social rate of time preference (i), the appreciation rate (a), the total rate of depreciation [including expected value of storm damages (m)], and the expected rate of reinvestment (h). Assuming that initial investment, depreciation, and reinvestment begin in year 0, and returns begin in year 1, then the marginal values of new investment and reinvestment can be expressed as:

Marginal value of new investment
$$\frac{\dfrac{r-a}{1+a} + m + h \cdot \dfrac{(r-a)}{(1+a)}}{\dfrac{i-a}{1+a} + m + h \cdot \dfrac{(i-a)}{(1+a)}} \text{ and}$$

Marginal value of reinvestment
$$\frac{\dfrac{(r-a)}{(1+a)} \cdot \dfrac{(i-a)}{(1+a)} + m \cdot \dfrac{(r-a)}{(1+a)} + \dfrac{r-a}{1+a} + m}{\dfrac{(i-a)}{(1+a)} \cdot \dfrac{(i-a)}{(1+a)} + m \cdot \dfrac{(i-a)}{(1+a)} + \dfrac{i-a}{1+a} + m}$$

As expected, when r equals i, the marginal values equal 1.0. When r exceeds i, the marginal values exceed 1.0. See Robert C. Lind, 1982, "A Primer on the Major Issues Relating to the Discount Rate for Evaluating National Energy Options," in R. C. Lind, ed., *Discounting for Time and Risk in Energy Policy*, Washington, D.C.: Resources for the Future, pp. 21-94.

13. Howard Kunreuther, 1978, *Disaster Insurance Protection: Public Policy Lessons*, New York: John Wiley & Sons.

14. People's perceptions are simulated to increase slowly over time at the rate at which the sea level rise signal could be statistically distinguished from the noise. At the beginning of the analysis, people are assumed to believe with equal probability that the sea level is either not rising or is rising at the rate of the recent historical trend for their location. As time goes on, their expectations gradually approach the true scenario (low, medium, or high) at the rate at which the lowest believed scenario can be statistically differentiated from the scenario being investigated. Statistical differentiation is assumed to be two standard deviations of annual average sea level position. This approach implicitly assumes that sea level rise will not change the variance of unrelated fluctuations in sea level position and that direct observation is the information most strongly affecting people's perceptions and hence behavior.

15. U.S. Army Corps of Engineers, Galveston District, 1979, *Galveston Study Segment: Texas Coast Hurricane Study, Feasibility Report*, Galveston: U.S. Army Corps of Engineers.

16. This question could be approached using regional economic models. The supply and demand links between the coastal and inland regions would be quantified, followed by an analysis of how changes in economic conditions in the coastal regions would affect the inland regions. A detailed transportation analysis may also be warranted.

17. In optimization programming, this is generally referred to as the specification of the objective function. In a complex problem such as identifying the optimal actions to take in anticipation of sea level rise, the objectives will be multi-dimensional and in a variety of units. Consequently, relative weights must be given to the various objectives reflecting value judgments. Additionally, the nonlinear and dynamic aspects of the analysis would require either unrealistic simplifying assumptions or considerable computing power.

18. Total damages will far exceed damages to privately owned buildings. Two estimates of damages from earthquakes put total damages at two to four times the damage to privately owned structures. A conservative estimate of twice the damage was used here. See Harold C. Cochrane, 1974, "Predicting the Economic Impact of Earthquake," in *Social Science Perspectives on the Coming San Francisco Earthquake: Economic Impact, Prediction, and Reconstruction*, Boulder: University of Colorado, and Worcester, Mass.: Clark University, p. 32.

19. Tax assessment maps and data were provided by Robert W. Ragin, assessor, Charleston County. The tax data provided by Mr. Ragin form the foundation upon which the economic analysis was built, and this project is indebted to him for his cooperation and assistance.

20. The storm surge information was obtained from FEMA Flood Insurance Rate Maps for the 100-year flood. Estimates of the 10-year and 50-year storm surge elevations were provided by RPI. The area loss for each sea level rise scenario for each subunit was developed by RPI.

21. The topographical data describing each subunit were developed by John Jenson, University of South Carolina.

22. The number of existing structures was obtained from property tax data supplied by Robert W. Ragin (see note 19). The average value of structures by land use within each subarea was computed by multiplying the assessed value of the structure by an empirically derived ratio of market value to assessed value, also provided by Mr. Ragin.

23. Land use changes and rates of growth were identified from planning documents from the city of Charleston, city of North Charleston, and Mount Pleasant.

24. Stephen Leatherman, Michael S. Kearney, and Beach Clow, 1983. *Assessment of Coastal Response to Projected Sea-Level Rise: Galveston Island and Bay, Texas,* URF Report TR 8301; report to ICF under contract to EPA, College Park: University of Maryland.

25. Ibid.

26. The distribution of topographical elevations above mean sea level within each subarea was developed at the University of Maryland by Stephen Leatherman and Beach Clow.

27. Carlton Ruch of the Research Center, College of Architecture, Texas A&M University, provided economic and population data developed in his ongoing research of the effects of hurricanes. The data provided by Dr. Ruch not only contributed significantly to the Galveston case study but also provided a model after which the development of the data for the Charleston case study was patterned.

28. Expected changes in land use were indicated in the data provided by Dr. Ruch (see note 27). These data were augmented with information in local planning documents from Texas City, Texas.

29. Storm damages were simulated by interpolating between three storm types (a 10-year storm, 50-year storm, and 100-year storm) to calculate a frequency-damage function that is integrated to estimate the expected value of damages in a given year. Protective structures (such as seawalls and levees) produce a discontinuous frequency-damage function. Although the continuous nature of sea level rise produces a continuous shifting of the discontinuity, the use of only three storm types to develop the frequency-damage function results in a large discontinuous jump in damages as soon as one of the three storm types overtops the protective structure. A more sophisticated model of the impact of protective structures on storm surges whose elevations exceed the height of the protective structures would eliminate this problem of discontinuity.

30. The mechanisms via which protected areas become vulnerable to storm surge in the high scenario are described in Leatherman et al. (see note 24).

Chapter 8

Planning for Sea Level Rise before and after a Coastal Disaster

James G. Titus

INTRODUCTION

Ocean beach resorts in the United States have always faced erosion and storm damage. At first, these risks were accepted as inevitable. Development was generally sparse, and people often built relatively inexpensive cottages along the ocean that they could afford to lose. When the occasional severe storm destroyed these houses and eroded the beach, replacement structures were frequently built farther inland to maintain the original distance from the shore.[1]

After World War II, beaches became more popular and were developed more densely than before. The resulting increases in real estate values enabled greater numbers of communities to justify expensive engineering solutions to maintain their shorelines. Frequently subsidized by the federal government, the practice of stabilizing shorelines replaced the previous custom of accepting erosion as inevitable.

The projected rise in sea level poses a fundamental question: how long should these communities hold back the sea? In the decades ahead, the costs of shoreline protection will rise dramatically and the relative efficiencies of various measures will change. But without such efforts, a 1 ft rise would erode most shorelines over 100 ft, threatening recreational use of both beaches and adjacent houses. Even under the low scenario, this could happen by 2025.

Although sea level is not expected to rise rapidly until after 2000, resort communities may have to consider its consequences much sooner. After the next major storm, in particular, homeowners whose properties are destroyed will decide whether and how to rebuild; and local governments will decide whether or not to let all of them rebuild, and which options are appropriate to address the storm-induced erosion. How well a community ultimately adapts to sea level rise will depend largely on the direction it takes when it reaches this crossroads.

This chapter examines the impact of sea level rise on the decisions that must be made before and after a coastal disaster. We first sketch the impact of sea level rise on coastal resorts, as well as the implications of recent federal policy changes. Using Sullivans Island, South Carolina (part of the Charleston study area) as an example, we discuss the impact of sea level rise on property owners' decisions on whether to rebuild if a storm happens to destroy their oceanfront houses in 1990. We then discuss the community's interest in this individual decision, as well as other decisions facing local governments. We conclude by discussing several policy changes that would enable coastal communities to better prepare for a rising sea.

THE CHANGING ENVIRONMENT OF COASTAL ACTIVITIES

A premise underlying most development in coastal areas has been that risks from storms and beach erosion as well as government responses to them will, on average, stay the same. However, federal policy changes and the prospect of sea level rise are destroying the validity of that assumption. If sea level rise is not adequately addressed, erosion may rob resorts of their recreational beaches and make oceanfront houses more vulnerable to damage from storm waves. In fact, there is some evidence that this is already happening (Pilkey et al., 1981; New Jersey, 1981; Massachusetts, 1981; U.S. Department of the Interior, 1983).

Many communities along the Atlantic and Gulf coasts are concerned about erosion[2] that is primarily driven by the current sea level rise of 1 ft per century (Pilkey et al., 1981; Bird, 1976).[3] Furthermore, some geologists note that since the last major storm along the Atlantic Coast (the 1963 northeaster), the underwater portion of many beaches has eroded much more quickly than the visible portion.[4] They argue that this phenomenon —known as "profile steepening"—implies that beaches would erode more quickly during a storm today than when their profiles were flatter and thus may no longer provide as much protection from storms as in the past. For example, Trident (1979) reported that at Ocean City, Maryland, the visible part of the beach is eroding 2 ft per year, but the underwater part is eroding 7 ft per year. As a result, Humphries et al. (1983) concluded that the beach would now protect structures for only one tidal cycle (12 hours) during a major storm and that even a 10-year storm could inflict considerable damage.

Sea level rise can change the effectiveness of engineering solutions to halt beach erosion. For example, the most common method has been constructing groins, which may curtail erosion caused by alongshore currents. However, they do not prevent sand from being carried offshore, which is the type of erosion caused by sea level rise (see Sorensen et al., Chapter 6). Thus groins have failed to stop erosion in many communities.

An ongoing realignment of the responsibilities of private property owners and federal, state, and local governments is also changing coastal development decisions. In the past, subsidized flood insurance has sometimes encouraged people to develop "high hazard" zones, transferring the economic risks to federal taxpayers. However, the federal Flood Insurance Administration intends to end these subsidies by 1988.[5] In the future, property owners will bear most of the costs from building in risky areas.

Furthermore, communities that benefit from projects that make areas less hazardous will probably have to pay more of the costs of these projects than in the past. After the 1900 hurricane killed 6,000 people in Galveston, the Army Corps of Engineers built a large seawall; and after the 1962 northeaster, the Corps supplied emergency sand to restore beaches. But the federal government is now less inclined to subsidize large-scale engineering projects. Budget-minded congressmen are less likely to vote for such projects, and when they do, they require substantial state and local contributions. States are also less likely to subsidize these projects than in the past.

THE INDIVIDUAL'S DECISION ON WHETHER TO REBUILD AFTER A STORM: SULLIVANS ISLAND

If a storm devastated a resort community today, the oceanfront houses that were destroyed would probably be rebuilt. To some people, the recreational value of being close to the beach justifies the risk of having their houses destroyed or, more recently, the cost of flood insurance premiums. If homeowners can continue to rely on the government to stabilize the shoreline, sea level rise may not substantially change this view. But if people expect their properties to be lost to erosion or the costs of maintaining them to increase (due to storm damage or higher insurance premiums), then rebuilding may be less attractive.

The case of Sullivans Island provides a conservative numerical illustration. This barrier island is typical of many family-oriented, noncommercial resort communities (see Kana et al., Chapter 4, for a description). However, the island's shoreline is currently advancing seaward, unlike most shorelines.[6] Kana et al. estimate that under current trends, the shore will advance 25-455 ft by 2025. Although the low scenario will cause parts of the shore to erode 45-70 ft, other parts will advance over 200 ft. Under the high scenario, the shore will erode 50-200 ft.

Gibbs (Chapter 7) divided the part of Sullivans Island within the case study area into 11 geographical zones. Table 8-1 shows his estimates of land values and structural values, as well as his estimates of the damages that would be sustained if a 100-year storm were to strike in 1990. It also shows the present value of the post-1990 damages caused by the low and high sea level rise scenarios.

Because much of the shoreline would still be advancing, the losses from sea level rise in the low scenario would be small (except in Zones 1 and 6). However, in the high scenario, the losses would be very significant in one-half of the zones. A 100-year storm in 1990 would destroy two-thirds of the value of the structures in Zones 1, 2, 6, and 7. Repairs would cost $27 million for the entire study area.

Given the immediate repair costs and future damages from sea level rise, property owners may want to reconsider whether the advantages of having an oceanfront house still justify the expense. Using data from Table 8-1, we calculated whether rebuilding makes economic sense for particular zones. (Data limitations forced us to conduct our analysis on entire zones and consider only the case where the homeowner rebuilds the original house.) We assumed that a rational property owner will rebuild only if the benefits of doing so exceed the costs. This will be the case only if the remaining value of a property is greater than the damage expected from sea level rise.

We calculated the remaining value of a property as the value of the land and structures (before sea level rise) minus the damage from the storm. Table 8-1 shows that in the case of Zone 1, this value is $2.72 million. Next, we compared this value with the damages from sea level rise. Table 8-2 shows that under the high scenario, the expected damage would be $2.99 million, and property owners would thus save $0.27 million by abandoning their properties instead of rebuilding.

Table 8-1. Estimates of Property Values and Expected Damages from Storms and Sea Level Rise (in millions of 1980 dollars)

Zone	(1) Value of Land	(2) Value of Structure	(3) Damage from a 100-Year Storm in 1990[a] (percent)	(1)+(2)−(3) Remaining Property Value after 100-Year Storm	Present Value of Estimated[b] Loss from Sea Level Rise	
					Low Scenario	High Scenario
1	1.27	4.23	2.78 (66)	2.72	1.11	2.99
2	0.99	3.32	2.13 (65)	2.17	0.42	1.67
3	1.23	4.11	2.38 (58)	2.96	0.43	1.45
4	1.26	4.22	2.06 (49)	3.42	0.34	0.83
5	0.32	1.06	0.42 (40)	0.96	0.07	0.13
6	2.27	7.60	5.00 (66)	4.86	1.69	5.13
7	1.33	4.45	2.86 (65)	2.91	0.56	2.22
8	1.76	5.89	3.47 (58)	4.24	0.61	2.01
9	2.79	9.33	4.56 (49)	7.56	0.76	1.75
10	0.83	2.78	1.10 (40)	2.51	0.20	0.34
11	0.27	1.92	0.26 (29)	1.93	0.06	0.09
Total	14.32	48.91	27.02 (55)	36.24	6.25	18.61

Source: Data from Chapter 7 (unreported result).
[a]Based on repair costs.
[b]Present-value based on a 5 percent real after-tax discount rate.

However, in the low scenario, property owners would lose $1.61 million more by abandoning their property than by rebuilding.

If property owners in Zone 1 could find someone to buy their land at its 1980 value, they would be less likely to rebuild. Table 8-2 shows that not rebuilding in the low scenario would yield a net loss of $0.33 million, and in the high scenario, a net gain of $1.55 million. The increased risks from sea level rise would discourage other families from buying the land to build a beach house. But someone with an alternative use in mind might be more interested in the property. For example, local governments might want to purchase land close to the shore in order to preserve a recreational beach.

To summarize these results: if property owners expect the high scenario, those in Zones 1, 2, 6, and 7 would be better off selling their property at the 1980 land value than rebuilding. Owners in Zones 1 and 6 would save money by not rebuilding their houses, even if they could not find a buyer and had to abandon their properties. Under the low scenario, rebuilding after a 100-year storm would be justified for all zones. However, if the houses in Zones 1 and 6 were entirely destroyed by a storm, property owners would save money by selling their land.[7]

Table 8-2. Analysis of Decision to Rebuild or Sell (in millions of 1980 Dollars)

Zone	(1) Remaining Property Value After 1990 Storm	(2) Projected Damage From SLR	(3) = (2)−(1) Net Savings From Abandonment	(4) Decision to Abandon or Rebuild	(5) 1980 Land Value	(6) = (3)+(5) Net Savings From Selling Out	(7) Decision to Sell or Rebuild
			If High Scenario Is True				
1	2.72	2.99	0.27	Abandon	1.27	1.54	Sell
2	2.17	1.67	−0.50	Rebuild	0.99	0.49	Sell
3	2.96	1.45	−1.51	Rebuild	1.23	−0.28	Rebuild
6	4.86	5.13	0.27	Abandon	2.27	2.54	Sell
7	2.91	2.22	−0.69	Rebuild	1.33	0.64	Sell
8	4.24	2.09	−2.15	Rebuild	1.76	−0.39	Rebuild
			If Low Scenario Is True				
1	2.72	1.11	−1.61	Rebuild	1.27	−0.34	Rebuild
2	2.17	0.42	−1.75	Rebuild	0.99	−0.76	Rebuild
3	2.96	0.43	−2.53	Rebuild	1.23	−1.30	Rebuild
6	4.86	1.69	−3.17	Rebuild	2.27	−0.90	Rebuild
7	2.91	0.56	−2.35	Rebuild	1.33	−1.02	Rebuild
8	4.24	0.66	−3.58	Rebuild	1.76	−1.82	Rebuild

Table 8-3. Costs of Incorrectly Anticipating Sea Level Rise When Rebuilding or Selling Out Are the Available Options (in millions of 1980 dollars)

Scenario Expected Actual Scenario	Low High	High Low
Zone		
1	1.54	0.34
2	0.49	0.76
3[a]	0	0
6	2.54	0.90
7	0.64	1.02
8[a]	0	0
Total	5.22	3.07

[a]Property owner would choose to rebuild regardless of the scenario and thus would not lose money by expecting the wrong scenario.

The Risks of Over- and Underestimating Sea Level Rise

Given the uncertainty about future sea level rise, property owners would have to weigh the costs of both overestimating and underestimating sea level rise. Table 8-3 illustrates the costs of being wrong for the six zones most affected by sea level rise.[8] (In this case we assume they have the option of selling out at the 1980 land value.)

If owners in Zone 1 assume that the high scenario is true, they will sell out (as shown in Table 8-2). But if the low scenario actually occurs, selling out will have cost them $0.34 million. On the other hand, if the property owners expect the low scenario when the high scenario is true, they will rebuild and forego the $1.54 million savings from selling out. Thus, unless they believe that the low scenario is five times more likely than the high scenario, property owners can reduce their expected losses by not rebuilding.

This type of analysis illustrates the benefits to property owners of securing better information. For several zones, the decision on whether to rebuild depends on which scenario is expected. Millions of dollars will be lost if property owners incorrectly project future damages from sea level rise. More certainty about whether the government will provide shore protection, as well as better forecasts of sea level rise, could enable property owners to avoid these losses.

Limitations of This Analysis

Three weaknesses limit the relevance of this numerical illustration. First, the only options considered were rebuilding the original house or not rebuilding at all. Other options might be preferable. For example, where abandonment is preferable to rebuilding the original house, building a cheaper structure might be even better. By ignoring potentially superior responses, our calculations may understate the benefits of planning for sea level rise.

Second, using zonal averages obscures impacts on individual properties and probably understates the impact of the low scenario on post-disaster decisions. For example, although the expected damage is a small portion of the entire value of Zone 1, Kana et al. (Chapter 4) project that in the low scenario, erosion would destroy some of the houses along the ocean by 2025. These houses would also be more vulnerable to a 100-year storm than the average house in their zone. Thus, some houses would probably not be rebuilt even if people expect the low scenario.

Finally, Sullivans Island is far less vulnerable to sea level rise than most coastal barriers. The starboard jetty at the entrance to Charleston harbor has modified ocean waves in a manner that causes the beach to advance along most of the Island. Thus, the impact of the high scenario on Sullivans Island would be comparable to the impact of the low scenario on most other barriers.

DECISIONS FACING THE COMMUNITY

The most important issue for resort communities to resolve will be whether to hold back the sea or retreat landward. The previous section assumed that the major impact of sea level rise on a homeowner's post-disaster decisions will be property losses from increased storm damage and erosion. However, public officials must also consider the impact of rebuilding oceanfront houses on the recreational use of the beach. The fact that a property owner might choose to rebuild his house in spite of projected erosion does not necessarily imply that the community's interest would be served by allowing the owner to do so.

Whether or not a community decides that a retreat is inevitable, the post-disaster period will be a critical time for implementing its response to sea level rise. If the community intends to defend its shoreline, it must do so soon after the storm, or redevelopment activities will be vulnerable to even a moderately severe storm. If it does not intend to fight erosion, deciding not to redevelop oceanfront lots can save the expense of later removing or protecting these properties. Finally, a major storm would increase the public's awareness of the consequences of sea level rise and thereby create a political climate more favorable to the difficult decisions that must be made.

Most measures by which resort communities can respond to sea level rise have been implemented or proposed in existing coastal hazard mitigation programs. These measures include building seawalls and other structures, pumping sand, restricting development or redevelopment, purchasing land, and modifying building codes and zoning.

Defending the Shoreline

The most commonly used measures to curtail erosion have been groins and beach nourishment. By groins, we mean long thin structures perpendicular to the shore that collect sand moving downshore, including jetties on the updrift side of inlets. By beach nourishment, we mean dredging sand from a channel or offshore and pumping it onto the beach (see Chapter 6).

Groins cannot prevent erosion caused by sea level rise, but they can move the problem downshore. A jetty at the south end of Ocean City, Maryland (acting as a long groin) has collected enough sediment to allow the shore to advance hundreds of feet, while to the south, Assateague Island National Seashore is eroding rapidly. As sea level rises, communities may use increasingly sophisticated methods to trap sand as it moves along the shore, in spite of the problems these measures may cause their neighbors.[9] In contrast, beach nourishment does not adversely affect neighboring areas, although it may be more expensive than groins.[10]

As Chapter 1 describes, beaches follow characteristic profiles. A 1 ft rise in sea level would eventually require raising the entire beach profile 1 ft. A profile that extended out to sea 0.5 mi would ultimately require 500,000 yd^3 of sand for every mile of beach. Estimates from Chapter 6 suggest that this would cost $2-5 million. In many resorts, the value of the property that would be protected could justify this level of expenditure.

However, sand pumping costs could vary considerably. Profiles extend to sea by very different amounts. For example, the profile of San Francisco may extend out several times farther than most profiles (U.S. Army Corps of Engineers, 1979), implying that protection costs would be several times greater. The availability of sand also varies considerably. Finally, the costs of beach nourishment may escalate as inexpensive supplies are exhausted. In spite of these uncertainties, such extremely valuable real estate as Miami Beach and Atlantic City could probably justify the costs in any event.

Communities that could not afford to raise their entire beach profiles might still use beach nourishment as a temporary measure until depreciation of oceanfront development or a storm makes retreat economical. As Leatherman points out in Chapter 5, much of the erosion from sea level rise does not take place until a major storm arrives. Until then, a small portion of the sand ultimately required to raise the profile may be sufficient to expand the recreational beach and provide useful protection from moderate storms. Even the erosion caused by a 100-year storm involves only a fraction of the sand required to raise an entire beach profile by 1 ft. For example, the Army Corps of Engineers' (1980) plan to protect Ocean City, Maryland from a 100-year storm would require only 150,000 yd^3 of sand per mile of beach, at a cost of less than $1 million/mi.

Planning a Retreat

Communities that decide to migrate landward could combine engineering and planning measures. One possible engineering response for barrier islands would be to preserve their total acreage by pumping sand to their bayside, imitating the natural overwash process.[11] This option might require bayfront property owners to be compensated for their loss of access to the water. Furthermore, care would have to be taken to ensure that marine life was not irreparably damaged. But because less sand would be necessary, such a program would be less expensive than pumping sand to the oceanside. In the long run, it would probably be less environmentally disruptive than any of the alternatives, particularly if mainland marshes are also allowed to migrate landward.

Planning measures will be important. North Carolina already requires most new home construction to be set back from the shore a distance equal to 30 years of erosion.[12] For existing construction, communities could implement strong post-disaster plans. Humphries et al. (1983) recommend that Ocean City, Maryland impose a temporary building moratorium after a major storm to give authorities time to decide which redevelopment is appropriate. However, the need to repair damages quickly may inhibit the careful debate necessary to adequately consider sea level rise.

Although many post-disaster development decisions cannot be made until local officials assess the damages, the general principles of redevelopment should probably be decided in advance of a storm. An assemblyman once introduced a bill to the New Jersey legislature that would have forbidden people to rebuild oceanfront houses that were more than 50 percent destroyed by a storm (Assembly Bill 1825). That bill was extremely unpopular, in part because it made no provision to compensate property owners. Our analysis of individual decisions suggests that if sea level rise is anticipated, many property owners who are offered some compensation will be willing to sell their land and write off their partly damaged houses; some might even do so without compensation.

In some instances, public officials might have to resort to eminent domain to purchase oceanfront property. Partly because of flood mitigation programs that require houses to be built on pilings sunk far into the ground, erosion from sea level rise will not always destroy the oceanfront houses now being built. Instead, some houses will continue to stand on the beach and perhaps even in the water. Although the owners of these houses might not want to move, the obstruction of the beach might be intolerable to the community and hence necessitate purchases under eminent domain.

Reaching a Decision

Public officials can use the same type of analysis as individual property owners to select the best policy, but they must also convince the public that they have reached the correct decision. Until the general public is convinced of the validity of the sea level rise projections, officials on coastal barriers may have trouble adopting the necessary responses.

Nevertheless, these officials should not defer all action until a scientific consensus emerges. As shown in the previous section, property owners in a community could save millions of dollars if they could be certain about the government's intentions. For example, property owners might conclude that sea level rise will make their property too hazardous to rebuild. If the government would stabilize the shoreline in the face of sea level rise, then announcing this policy in advance would enable these people to enjoy their properties rather than mistakenly assume that sea level rise threatens them.

Deciding on the best response to sea level rise could take communities many years. By the time this process is complete, better forecasts of sea level rise may be available. Because we cannot know when a major storm will occur, the time saved by initiating the planning process sooner rather than later may be the critical difference between being ready to act and being unprepared.

RECOMMENDATIONS

In the next few decades, sea level rise will force resort communities to either retreat several hundred feet or defend their shorelines by pumping sand. Where sand is inexpensive or real estate extremely valuable, it may be desirable to raise an entire barrier island and associated marshes in place. But if the shore retreats during a storm that destroys oceanfront structures, reclaiming the land lost would accomplish much less at a higher cost and thus be more difficult to justify.

Recent experience with post-disaster development suggests a strong inclination for homeowners to rebuild. However, subsidized beach nourishment and insurance have insulated people from the consequences of building in hazardous areas, and increasing risks from sea level rise have not been expected. As these conditions change, the relative merits of defending the shore versus migrating landward will change as well.

The political climate is rarely receptive to policies that impose costs now to protect against unknown risks in the future. But that climate will never be more favorable than when people are in the midst of recovering from a disaster that could have been avoided. Outlays for land purchases or beach nourishment would necessitate large increases in property taxes. But the drop in property values that might ultimately occur otherwise could be even greater.

On the basis of the information in this book, government officials should consider the following recommendations:[13]

1. *Post-disaster plans should determine which policies are appropriate if sea level rise is expected.* Because successful crisis management requires planning in advance of the crisis, many states are developing post-disaster plans for coastal communities. These plans would make excellent vehicles for preparing for sea level rise.

2. *Local governments should inform the public of the risks from sea level rise and begin to formulate responses.* This policy could enable property owners to avoid losses from uncertainty. Our analysis of Sullivans Island indicates that if sea level rise is expected, many property owners would base their decision to rebuild on whether the government was going to stabilize the shoreline. If the willingness of a community to pump sand is limited, a retreat of the shore is inevitable. Waiting until oceanfront lots are redeveloped to confront this issue would substantially increase private losses as well as the costs to communities of purchasing oceanfront property and otherwise adapting to sea level rise.

3. *State and local governments should determine research needs and inform policy makers and research institutions of these needs.* Coastal communities represent one of many constituencies with particular research needs. Only if they make their needs known is the necessary research likely to take place.

4. *All beach nourishment efforts should consider the sand required in the long run as well as the measures that are necessary to maintain the visible portion of the beach.* The steepening beach profiles that already worry many coastal geologists could be exacerbated by short-sighted policies to stabilize shorelines, particularly if sea level rise is not recognized to be causing the erosion.

5. *Policies that prohibit bayside filling should be modified to permit landward migration of developed barrier islands in step with oceanside erosion and the migration of undeveloped barriers.* Policies that prevent

local interests from creating new acreage by filling the baysides of barrier islands have been necessary to preserve important marine ecosystems. But as the sea rises, an inflexible adherence to this principle could harm the very environments these policies seek to protect.

In the last several thousand years, marine life adapted to the landward migration of barrier island ecosystems as the sea rose. If developed barriers are prevented from migrating, while undeveloped barriers migrate, there is no guarantee that these species would do as well: inlets would widen, changing the tides and increasing the salinity of the bays considerably; estuaries would deepen and perhaps deprive marshes of necessary sediment. Furthermore, the engineering alternatives to landward migration would disrupt oceanside environments. The impacts of preventing migration could easily be more disruptive than filling 100 feet of an estuary every few decades, provided that the estuary and bayside marshes also migrate landward.

The potential savings in sand pumping costs will not always outweigh the institutional problems of entire communities migrating landward. But where they do, environmental policies should not prohibit sincere efforts by engineers to decrease human interference with natural processes.

6. *Interpretations of riparian rights should recognize the scarcity of sand and should treat interference with the natural transport of sediment the way they treat interference with the natural flow of water.* Only time will tell whether beach nourishment or groins develop sufficiently to become the least-cost supply of sand. But if groins prove to be the cheaper, then more controls on their use will be necessary. Otherwise, coastal communities may waste considerable resources fighting over limited quantities of sand, and undeveloped barrier beaches will be effectively "mined" of their sustenance.

This possibility is especially likely for communities next to inlets where navigation requires jetties, such as Sullivans Island and Ocean City. A jetty will collect sand up to a point, after which the excess sand washes downdrift again. In the face of a rising sea, even the most enlightened local leadership will be hard pressed to explain why this excess should not be pumped onto the rest of the community's beaches but must instead be allowed to wash away.

When sea level was assumed to be stable and erosion was thought to result mainly from alongshore currents, riparian rights correctly did not interfere with communities' attempts to build groins and "keep their own sand." However, as sea level is recognized to cause the erosion, attempts to halt it by capturing sediment moving alongshore can only be interpreted as attempts to divert sand from its natural destination, the beaches downshore. These diversions will sometimes be justified, as dams for irrigation are sometimes justified. But justice will be best served if riparian rights come to recognize the increasing scarcity of sand, as they have with water.

7. *The Federal Emergency Management Agency should develop policies that encourage communities to address shoreline retreat.* The National Flood Insurance Program was designed to ensure that future development did not create the conditions that make coastal disasters likely. Building codes now require most new construction to be elevated above flood levels. Furthermore, the expected end of insurance subsidies will ensure that the only people to build in hazardous areas will be those who can afford to pay for the expected damages. However, the program has not been given a mandate to prevent disasters caused by shoreline retreat. Other than raising insurance premiums on properties that erosion makes more vulnerable to storms, the program does not address erosion.

By creating the Flood Insurance Program, Congress determined that the importance of preventing future coastal storm disasters transcended the laissez-faire notion that the marketplace adequately acknowledges risks from storms. This reasoning applies equally to coastal disasters caused by erosion and sea level rise. Communities devastated by storms will receive federal disaster assistance. In providing this assistance, the Federal Emergency Management Agency should encourage these communities to prepare for a rising sea. The improvements in technical and institutional capabilities that result from their early experiences could repay society many times over in the years to come.

8. *Coastal barrier communities should consider impacts on marshes, estuaries, and mainland activities when planning for sea level rise.* This chapter has discussed whether coastal barrier communities should defend their shores or migrate landward, from the narrow perspective of barrier residents and economic interests. However, the actual decision process will also involve representatives concerned with marshes, estuaries, and the mainland. We believe that in most cases, either the barrier and the mainland shore will both retreat landward, or they will both be stabilized. (Otherwise, the environmental and navigational impacts on estuaries and marshes would be severe, and perhaps intolerable.) Therefore, coastal barrier communities must take an interest in the fate of the adjacent mainland.

The impacts of sea level rise will confront coastal barriers sooner than the mainland. Nevertheless, planning the mainland's future will require more lead time because the decisions involved will be more controversial. While barrier communities may decide to migrate landward on purely economic grounds, a major reason for mainland development to retreat will be the environmental impact of not allowing marine ecosystems to adapt to sea level rise. Landward migration will require coastal barriers only to yield storm-damaged structures to an ocean that would have been difficult to hold off anyway. In contrast, a retreat will require the mainland to yield both land and useful structures that could be inexpensively protected from encroaching marshes and estuaries.

Some people argue that it is too soon to prepare for sea level rise because the phenomenon is so far in the future and has not yet been proven. Paradoxically, this situation may make some types of farsighted decisions easier to implement. For example, zoning could require that particular properties be abandoned in the next century if sea level rises a certain number of feet. Such a measure would appeal to environmentalists who want to ensure that coastal barriers and marine ecosystems will be allowed to adapt naturally to sea level rise. Property owners who do not care about the next century or do not believe that the sea will rise should not object to such measures.

If local governments implement this type of planning measure, the necessary abandonment of low-lying areas can be conducted in an orderly fashion. In this way, our generation can save future generations from costly clashes between competing economic and environmental interests. But even if objections are raised, at least we will learn that the public already understands the prospect of sea level rise and cares about the next century more than we realized, and that would be a start.

NOTES

1. A walk along the beach at Ocean City, Maryland, for example, reveals evidence of past shoreline retreat. For the most part, the front row of houses is behind the vegetation line, but an occasional house can be found standing closer to the water, a survivor from a previous generation of development. These houses generally stand on what is now public land and cannot be rebuilt if they are destroyed by a storm.
2. See Taylor, Ronald A., "America's Losing Battle to Save Its Beaches," *U.S. News and World Report*, July 11, 1983, pp.51-52.
3. The Bruun Rule would predict that the shore should be eroding several feet per year. See Chapter 1 for an explanation of how sea level rise causes erosion.
4. Conversations with Orrin Pilkey, Geology Department, Duke University; Stephen Leatherman, University of Maryland, and Stanley Humphries, IEP, Wayland, Massachusetts.
5. Federal Emergency Management Agency, *Justification of Program Estimates for FY '84*, February 1, 1983.
6. The island has a substantial supply of sediment and a jetty offshore that cause waves to retain much of it.
7. Given our simplifying assumptions, this conclusion automatically applies to all zones. The only remaining investment is in the land, and we assumed that property owners will receive the pre-sea level rise price, even though it would be worth less once people expect sea level rise. However, for Zones 1 and 6, the drop

in property values from sea level rise probably overshadows any possible errors from our simplifying assumptions. For example, in Zone 1, sea level rise would lower the value of the land to homeowners from $1.27 million to $0.16 million. An offer of the original price, or even somewhat less, would be difficult to refuse.

8. The cost of expecting the high scenario when the low scenario is true was calculated as the remaining value of the property, minus the damage from sea level rise under the low scenario, minus the land value (which is recovered from the sale). The cost of expecting the low scenario (and rebuilding) when the high scenario is true was calculated as the remaining value of the property, minus the losses from sea level rise (high scenario), plus the cost of rebuilding, minus the value of the land (which again is recovered from selling the land).

9. For example, the jetty at Ocean City allowed the southern mile of beach to advance considerably but now is "filled up" and allows sand to pass downshore. At the same time, the northern 8 mi of Ocean City are eroding. Residents at the northern end advocate pumping the excess sand near the jetty onto the rest of the beach. Such a plan might enable Ocean City to hold back the sea indefinitely.

10. In fact, one reason that this solution to erosion is not used more extensively is the public's perception that the sand will wash away and benefit neighboring communities as much as themselves. In contrast, groins are viewed as more permanent. Earl Bradley and Chris Zabawa, Coastal Resources Division, Maryland Department of Natural Resources, personal communication, July 1983.

11. See Chapter 1 for a discussion of overwash processes.

12. *North Carolina Administrative Code*, Chapter 7H, 1983. Raleigh, North Carolina Office of Coastal Management.

13. Policies appropriate for Louisiana would require a reconsideration of the costs and benefits of alternative levee, dredging, and water diversion policies. Such assessments are probably even more urgent than the issues discussed here. See the comments of Sherwood Gagliano, Chapter 10, pp. 296-300.

REFERENCES

Bird, E. C. F. 1976. "Shoreline Changes during the Past Century." In *Proceedings of the 23rd International Geographical Congress*. Moscow: IGC.

Humphries, Stanley, Larry Johnston, and Stephen Leatherman. 1984 (in press). *Reducing Flood Damage Potential at Ocean City, Maryland*. Annapolis: Division of Coastal Resources, Department of Natural Resources.

Massachusetts, State of. 1981. *Barrier Beaches: A Few Questions Answered*. Boston: Massachusetts Office of Coastal Zone Management.

New Jersey Assembly Bill 1825. June 1980.

New Jersey Department of Environmental Protection. 1981. *New Jersey Shore Protection Master Plan*. Trenton: Division of Coastal Resources.

Pilkey, O., J. Howard, B. Brenninkmeyer, R. Frey, A. Hine, J. Kraft, R. Morton, D. Nummedal, and H. Wanless. 1981. *Saving the American Beach: A Position Paper by Concerned Coastal Geologists*. Results of the Skidaway Institute of Oceanography Conference on America's Eroding Shoreline. Savannah: Skidaway Institute of Oceanography.

Trident Engineering Associates. 1979. *The Trident Report on Interim Beach Maintenance at Ocean City*. Annapolis: Division of Coastal Resources, Maryland Department of Natural Resources.

U.S. Army Corps of Engineers. 1979. *Ocean Beach Study: Feasibility Report.* San Francisco: San Francisco District.

U.S. Army Corps of Engineers. 1980. *Feasibility Report and Final Environmental Impact Statement.* Baltimore: Baltimore District, Atlantic Coast of Maryland and Assateague Island, Virginia.

U.S. Department of the Interior. 1983. *Undeveloped Coastal Barriers: Final Environmental Impact Statement.* Washington, D.C.: Coastal Barriers Task Force, DOI.

Chapter 9

Implications of Sea Level Rise for Hazardous Waste Sites in Coastal Floodplains

Timothy J. Flynn, Stuart G. Walesh,
James G. Titus, and Michael C. Barth

INTRODUCTION

On the night of July 20, 1977, 30 cm (1 ft) of rain fell on Johns-town, Pennsylvania, in a period of six hours. As a result, the Connmar River, which runs through the heart of the city's industrial district, overflowed its banks. Cylinders containing compressed gases, drums containing toxic chemicals, oil-soaked debris, and other hazardous materials were washed downstream and deposited on recreational and residential properties when the floodwaters receded.

Recognizing the potential threat to public health and the environment, the federal and Pennsylvania state governments immediately set up a joint task force to address this threat. At considerable cost, the clean-up team surveyed the area, collected the containers, analyzed their contents, and returned them to their owners or safely disposed of them. Although it would have been infeasible to locate and identify every container that washed away, about 500 cyclinders and 500 drums were collected in this effort.

Although there are 1,100 active[1] hazardous waste sites within 100-year floodplains in the United States (DPRA, 1982) and possibly as many closed or abandoned sites, flooding disasters such as the Johnstown incident have been infrequent in the past. However, a rise in sea level could significantly increase the probability of flooding for many of these sites and bring more sites into floodplains. Furthermore, erosion and salt intrusion that would result from a rise in sea level could become additional threats, even to those waste sites that are adequately protected against flooding.

This chapter first discusses the hazards associated with waste sites in floodplains and federal regulations to mitigate those hazards. It then discusses the potential impacts of sea level rise on specific types of hazardous waste sites and illustrates how sites in the Charleston and Galveston areas would become vulnerable to flooding. Finally, it presents the authors' conclusion that compliance with existing regulations could prevent serious problems with operating sites but not with closed or abandoned sites.

BACKGROUND

This section provides general information on the federal regulations and guidelines for the siting of hazardous waste facilities, particularly as they apply to floodplains. Then, the method of delineating the 100-year flood boundary and the general characteristics of floods are discussed.

Regulations for Siting Hazardous Waste Facilities

The Resource Conservation and Recovery Act[2] (RCRA) created a legal mechanism for the management of hazardous wastes. EPA implemented RCRA through a series of regulations and guidelines; a significant part of the regulations restrict waste disposal in environmentally sensitive areas. The following RCRA regulations[3] pertain to hazardous waste facilities in floodplains:

Floodplains: [B 264.18(b)] Hazardous waste surface impoundments, waste piles, land treatment units, and landfills preferably should not be located in a 100-year floodplain. Facilities so located must be designed, constructed, operated, and maintained to prevent washout of any hazardous waste by a 100-year flood. However. in accordance with §264.18(b)(1)(i), if the owner or operator demonstrates that, in the event of a flood, the waste would be removed to a safe area before flood waters reached the facility, special design and operating procedures to prevent washout are not required. This option may not be viable for many existing surface impoundments, waste piles, land treatment units, and landfills. Accordingly, the Agency is promulgating a second exemption, defining narrow circumstances in which existing facilities, not designed and operated to prevent washout, may be located in a 100-year floodplain without the owner or operator making the demonstration cited above in §264.18(b)(1)(i). These circumstances are where the owner or operator demonstrates that a washout would cause no adverse effects on human health or development. (CFR, 1982, *Federal Register*, vol. 47(143):32290-32291.)

EPA did not prohibit the operation of facilities in the 100-year floodplain, both because of the potential economic impacts of such a requirement and because of the availability of techniques to protect the facilities from floods.

EPA has adopted the Flood Insurance Administration's (FIA) minimum requirements[4] for new construction and substantial improvements of any nonresidential structure in 100-year floodplains. The requirements specify that the lowest floor of a building, including the basement, must either be elevated to the 100-year flood elevation or be floodproofed so that the structure is watertight and capable of withstanding the forces exerted by floodwaters during a 100-year storm. The standard specifies the same operational and design standards for new and existing facilities.

Methods of Delineating the 100-Year Flood Boundary

The FIA prepares flood insurance maps for communities subject to periodic flooding. Approximately 17,000 communities and counties are in the flood insurance program; however, the mapping coverage varies (FEMA, 1978). A community applying to enter the program is initially placed in the "Emergency Program" until certain requirements are fulfilled and then it enters the "Regular Program."

Flood hazard boundary maps (FHBM) are prepared for communities in the Emergency Program and have been prepared for nearly all flood-prone communities in the United States (FEMA, 1978). These maps show the areas within the community that are vulnerable to a 100-year flood. The floodplain boundaries are only approximate on these initial maps, and surge elevations are not provided. Owners or operators of waste sites may appeal the boundaries of the 100-year floodplain on a FHBM by submitting conclusive scientific evidence that the boundary is incorrect.

While a community is in the Emergency Program, flood insurance studies involving detailed field engineering surveys are conducted. The information obtained from these studies is used to prepare a flood insurance rate map (FIRM) and a flood boundary and floodway map. Because the FIRM delineates precise boundaries of the 100-year floodplain, it is the final determination of whether the facility is located within the 100-year floodplain.

Flood Characteristics

The most important flood characteristics are depth, velocity, duration, and fall.[5] The depth of floodwater around a structure is the most critical element in planning and designing flood control measures to meet strength and stability requirements. The velocity of floodwaters during overbank/ inland flow conditions has important implications for scouring, sediment transport, debris loading, and dynamic loading on structures and obstructions. The flood's duration determines the extent of saturation of soils and building materials, seepage, and water pressure in soils and under foundations. The rate of rise and fall of a flood to and from its crest affects the size of flooding and the required drainage provisions.

Advance warning from flood forecasting is important, particularly where flood control plans require time for wastes to be removed before floodwaters reach the facility. Floating debris can cause increased loads against structures, resulting in damage. Wave action caused by coastal storm surge can also result in significant hydrodynamic loads on structures in the flood zone. All these aspects of floods should be evaluated and their probable effects considered when siting and designing a hazardous waste facility in a floodplain.

HAZARDOUS WASTE FACILITIES IN FLOODPLAINS

EPA is preparing a regulatory impact analysis (RIA) for operating hazardous waste facilities located in or proposed for location in floodplains. Included in the preparation of the RIA are visits to hazardous waste facilities. This section discusses the preliminary findings of Walesh and Raasch (1983), who visited approximately 10 hazardous waste facilities during 1982 and 1983 as a part of EPA's assessment of the impact of the regulations discussed in the previous section. By including a wide range of waste management technologies and flood mitigation measures, the site visits provided a wealth of field-based ideas, information, and a representative view of the current situation. Site-specific field data combined with the flood mitigation and hazardous waste facility experience of the project team were the basis for the following observations. Although these observations do not represent a scientific sample, they do represent perhaps the only available systematic review of the flood mitigation experience of hazardous waste facilities.

Hazardous Waste Management Technologies and Flood Mitigation Options

Walesh and Raasch observed several waste management technologies, the most common being surface impoundments, containers (primarily 55 gal drums), and above-ground tanks. Most above-ground tanks were used to store hazardous wastes temporarily, and some were used to pre-treat wastes using processes such as solids settling and acid neutralization. The wastes handled by these sites included acids, alkalis, solvents, heavy metals, grease and oils, paint waste, and PCBs.

Flood mitigation options for hazardous waste facilities can be divided into two categories: flood protection (not allowing floodwaters to reach the facility) and floodproofing (allowing floodwaters to come into contact with structures but preventing any damage). The structural components of facilities such as tanks, containers, incinerators, and structures used in thermal, chemical, physical, and biological treatment can be floodproofed (EPA, December 30, 1980). Landfills, surface impoundments, land treatment, and waste piles require flood protection. The flood mitigation measures commonly being applied to hazardous waste facilities are the same as those commonly used to flood-protect or floodproof facilities and structures that do not handle hazardous wastes.

Because new facilities have more flood mitigation options available than existing facilities, the cost of achieving RCRA floodplain standards for them is generally less. Risks for new facilities (particularly those susceptible to inundation and severe wave action) can be reduced by considering not only the direction of the shore but also the orientation of the units in light of the probability of flooding from various directions. The proper orientation of structures may reduce impacts from receding floodwaters as well. Floodwater elevations and velocities associated with the 100-year storm must be identified at the facility's location to ensure the design of adequate flood control measures. Existing facilities located in 100-year floodplains must either modify management units at the facility, divert or prevent the entry of floodwater (e.g., levees), or relocate.

Flood Protection. Levees, generally earth mounds with trapezoidal cross-sections, are the most common means of flood protection. Walesh and Raasch observed an earth levee along the entire bay boundary of a large sanitary landfill on San Francisco Bay that contained a hazardous waste site. The portion of this levee exposed to wave action was armored with riprap. Another example was an earth levee along the San Joaquin River in California that both contained and flood-protected a series of surface impoundments constructed on a riverine floodplain at a chemical manufacturing plant.

Floodwalls, usually constructed of concrete or sheet steel and functioning like earth levees, are used where wave attack is significant or space is limited. However, they are not used as much as levees, probably because of their higher costs and the general availability of the larger space required for levees.

Hazardous waste facilities in coastal floodplains are often traversed by small creeks or drainage ways. The stormwater runoff these streams carry may significantly exacerbate the flooding from storm surge alone. Safely and economically accommodating this drainage is a major problem at some facilities. The owners of a large landfill in California, for example, plan to increase the capacity of two existing drainage ways so as to divert drainage around their hazardous waste management areas. To prevent interference with operations, street improvements and a large stormwater detention facility were recently constructed at the naval base in Charleston to capture and temporarily store drainage from higher land outside of the floodplain with subsequent slow release to the river. The owner of a manufacturing plant along the Taunton River in Massachusetts installed a floodproofing gate on an entrance to his plant. When floods are expected, the gate is closed to prevent drainage from entering the plant.

Floodproofing. The most commonly used methods of floodproofing are grading, fencing, and upgrading the structural integrity of containers; elevating containers and/or facilities above flood levels; and emergency plans to evacuate wastes. The elevation of storage units, incinerators, and treatment facilities on earth fill above flood levels is often effective, provided that the fill does not easily erode. The selection and placement of fill should be based on the effects of saturation from floodwaters, slope stability, scour potential, and on whether settlement is uniform or differential. Column piers or walls can also be used, provided they do not restrict the flow of floodwaters and are protected against scour.

Grading of landfills and land treatment areas is used to establish a slope for runoff. The slopes are generally greater than 2 percent to provide adequate drainage and prevent wastes from leaching but less than 5 percent to reduce flow velocities and minimize erosion (U.S. EPA, September 1980). Fencing that does not impede the flow of floodwaters has been installed around facilities to restrict flotation of containers and reduce structural damage from flood debris impact. The structural integrity of treatment units, storage facilities, and incinerators should be designed to withstand hydrostatic and hydrodynamic forces associated with the 100-year flood (Department of the Army, 1972).

A commonly used floodproofing measure is placing hazardous waste containers in locations above the flood level. Examples include a fenced drum storage area at a chemical manufacturing plant positioned near, but outside and above, the floodplain. Another observed example was a manufacturing plant located entirely within the Taunton River floodplain, with a hazardous waste drum and storage area on an upper floor of the building well above flood stage. The naval facility in Charleston plans to floodproof its new hazardous waste management storage building, which will be in a coastal floodplain, by constructing the entire facility on fill so as to elevate it above flood level.

Another common means of floodproofing is the development of a formal contingency plan or emergency action plan. A plan prepared by the naval facility includes preparation for emergency situations such as forecast hurricanes. A large chemical company in New Jersey and a manufacturing plant in Massachusetts also have plans that include temporarily removing drums of hazardous waste from flood-prone storage areas.

Flood Protection and Floodproofing Design. Professional engineers are responsible for the design and construction of the flood mitigation measures at hazardous waste sites. These measures are similar to those for other private and publicly owned structures and facilities located on floodplains.

Although there is general agreement that flood impact mitigation is an element of proper management at hazardous waste sites, there is no common approach used to select the design flood, which can be attributed primarily to two factors. First, there are widely differing perceptions of the flood threat and its economic, public health, and safety consequences relative to the cost of flood mitigation at facilities. For example, some facility managers view the 100-year recurrence interval flood as such a remote event that they cannot justify protecting facilities against it. However, protection designed against the 100-year storm is becoming the norm in most areas. Second, some existing facilities were built many years prior to the institution of hazardous waste management regulations or even state and federal floodplain management regulations. Therefore, facility designers used the design flood criterion they considered most economic and otherwise appropriate. For example, a chemical manufacturing facility in California used the largest flood on record to design berms to flood protect surface lagoons located on the San Joaquin River floodplain.

Nor has the concept of freeboard (extra height in a facility as a safety factor) been universally accepted and applied. In some cases, no freeboard is provided. When freeboard is incorporated in the design of flood mitigation measures, it is generally 1 m or less (2-3 ft).

Many operators of existing hazardous waste facilities would resist retrofitting existing facilities to provide flood mitigation for 100-year floods because they do not believe the costs are justified. Particularly in the case of older, congested facilities, flood mitigation retrofitting measures could be extremely disruptive and very costly.

The managers of new facilities appear much more likely to design for a 100-year flood. They would generally accept designing flood mitigation measures for the 100-year or similar flood, including a sea level rise increment, as a condition of occupancy of the floodplain or they may even accept simply staying out of the floodplain. For example, one operator, after strongly stating his opposition to imposing 100-year recurrence interval flood mitigation criteria on his company's existing hazardous waste and related facilities, told Walesh and Raasch that if his company were constructing a new manufacturing facility with its attendant hazardous waste generating, treating, and storage components, it would simply stay out of the floodplain.

Flood mitigation practices do not generally include secondary measures (redundancy) to offset the possibility of structural failure. For example, when a levee is used to protect a riverine community from flooding, buildings are not normally floodproofed to provide protection in the event that the levee fails. However, a variation on this concept is occurring at some hazardous waste sites in the form of written contingency plans or emergency action plans in the event that a levee, floodwall, or other structural flood mitigation measure appeared to be in danger of failure. These plans typically include evacuation of employees and temporary transport of hazardous materials to safe locations.

One of the coastal sites Walesh and Raasch visited provided a rare example of structural redundancy. A surface lagoon containing hazardous wastes at this privately owned landfill was encircled by a berm. The next level of protection was a berm containing a cut-off wall with the bottom of the cut-off wall keyed into relatively impermeable subsoil. Although the primary purpose of this combination was to contain liquid hazardous wastes that may discharge from the lagoon into the surrounding soil, the berm/cut-off wall combination could also serve as a flood protection measure. Beyond the berm/cut-off wall, the entire outer perimeter of the hazardous waste area, plus the adjacent landfill, was protected from high water and waves with a levee.

Implementing Flood Protection and Floodproofing Measures

In many cases, the primary motivation for the implementation of flood mitigation measures is to permit the continuous and safe operation of the entire facility as well as the protection of the public from toxic wastes. For example, the management of a large, privately owned and operated landfill with a hazardous waste area explained that they could not afford to have all or a portion of the facility temporarily closed by flooding. Other examples Walesh and Raasch identified included a manufacturing plant and a naval facility that generated hazardous wastes as part of their overall operations, with treatment and temporary storage of hazardous wastes being secondary activities. The managers of these two facilities explained that flood protecting and floodproofing the hazardous waste handling areas, as well as other vulnerable portions of their physical plants, were necessary so that flooding would not endanger employees and interrupt their major functions.

Rarely do facility managers indicate that local, state, or federal regulations were the primary motivation for instituting flood mitigation measures. However, the degree of protection provided is driven primarily by regulations. Although facility managers generally agree that flood mitigation measures are good business and management practices, they do not necessarily agree that protection should be provided for flood events as rare as the 100-year recurrence interval flood required by EPA regulations.

Costs of Flood Protection and Floodproofing Measures

The costs of hazardous material management (including, but not limited to, flood mitigation measures) can be high, particularly for retrofitting existing facilities. For example, a government facility Walesh and Raasch visited plans to construct a $1 million hazardous waste facility with only approximately 5-10 percent of the cost being used for floodproofing, that is, elevation of the facility on fill so that it will be above flood levels. However, it is usually difficult, if not impossible, for an operator to separate the flood mitigation costs of an existing or proposed facility from the other hazardous waste or even hazardous material handling costs. For example, a berm or wall constructed around a hazardous waste storage tank located in a floodplain may contain the wastes, as well as provide floodproofing for them.

Inspection and Maintenance of Flood Mitigation Measures

Inspection of flood mitigation measures at hazardous waste facilities is usually part of routine production operations. Inspection procedures were found to include daily inspection by hazardous waste personnel of all storage locations on a naval facility, periodic monitoring of impoundment embankments and groundwater quality by engineers, and essentially continuous inspection of drum storage areas by the personnel consolidating wastes and sealing drums. Maintenance of flood mitigation works is provided on an as-needed basis at most sites.

How Facility Operators View Environmental Protection

Most owners and operators of hazardous waste sites generally view treating, storing and disposing of hazardous wastes as a secondary activity—part of the cost of running their business. Examples are provided by the several manufacturing firms visited during the project as well as by government facilities that handle hazardous wastes. The most important exceptions are sanitary landfills that treat hazardous wastes, where the treatment, storage, and disposal of hazardous waste is one of the primary activities.

Although hazardous waste management is widely viewed as part of the cost of doing business, most operators believe that controls and their attendant costs are required to protect the environment, provided that the controls and costs are reasonable. According to one manager, his firm has significantly changed its attitude toward environmental matters in recent years and now openly and positively accepts its environmental responsibilities.

GENERAL IMPACTS OF SEA LEVEL RISE ON HAZARDOUS WASTE SITES

The impacts of sea level rise on hazardous waste sites can be classified into increased storm damage, shoreline retreat, and changes in water tables. Increased risk from storms is likely to be the most important factor. A rise in sea level would bring new sites into floodplains and result in more severe flood levels for those already in floodplains. Furthermore, the risks from damaging storm waves would increase as deeper water allowed these waves to penetrate further inland.

Shoreline retreat could also threaten hazardous waste sites. As Chapters 1, 4, and 5 explain, a sea level rise results in both inundation and erosion; a rise of 1 ft could result in shoreline retreat from a few feet along rocky coasts to several miles along low-lying marshland. Significant shoreline retreat might leave a waste site under water or in the surf zone subject to constant wave attack. Operators of existing sites, especially factories for which the waste site is a small portion of the entire operation, would generally protect their operation from an encroaching shoreline. Abandoned sites, however, would not be guaranteed the same protection.

Finally, changing water tables could threaten wastes stored in surface impoundments and landfills. Higher water tables could threaten containment vessels by exerting additional hydrostatic pressure. Furthermore, saltwater can permeate clay liners that are impervious to freshwater. As a result, the risk of wastes leaching through the liners would increase.

Although existing EPA regulations do not address changing water tables, the regulations would protect the public from risks associated with shoreline retreat and flooding for existing sites, provided that flood maps were redrawn in a timely fashion. However, if sea level rise is not anticipated and these sites then undertake the necessary mitigation actions to comply with EPA regulations, additional outlays that could have been avoided may be required to ensure protection against such a rise in sea level. Some facilities would have to redesign and rebuild their flood protection works to withstand the greater risks from sea level rise. Furthermore, other sites might have chosen to locate farther inland had they realized that their chosen site would someday be in a floodplain.

IMPACTS ON SPECIFIC TYPES OF HAZARDOUS WASTE SITES

Information on the potential impacts of coastal flooding on hazardous waste facilities is limited.[6] On the basis of this limited information, this section briefly describes the types of impact that sea level rise could have on landfills, land treatment areas, surface impoundments, waste piles, storage facilities, and incinerators.

The impacts of sea level rise on landfills are inundation, waste solution migration, physical erosion, and saltwater intrusion.

Inundation of a landfill can result if flood waters are high enough. A ponding effect will cause increased leachate production by adding water to the volume of wastes in the landfill and causing varying degrees of saturation (which may affect structural stability).

Waste solution due to floodwater may result in increased leachate production and the potential migration of these wastes onto neighboring properties. Active sites that are not capped are particularly vulnerable.

Waves can cause extensive erosion of any loose cover material. The degree of impact would relate directly to the amount of wave action resulting from a coastal flood. Erosion is particularly significant at landfills constructed so that the waste is above ground level.

Salt intrusion from sea level rise may affect landfills with clay caps and/or liners. In coastal areas, where the extent of saltwater intrusion inland may be significant, it is common to have shallow unconfined aquifers with depths that respond rapidly to fluctuations in sea level. A rise in sea level would result in a rise in the water table. The liner of a landfill may become inundated as the shallow water table rises, thus building up increased hydrostatic pressure on the liner. If the aquifer is contaminated with saltwater, there may be significant clay-salt interaction, which can result in increased permeability of the clay liner and potential migration of leachate from the facility. Recent research efforts have shown that increased salt concentrations may cause a decrease in the shear strength of clay, thus weakening its structural stability. Sodium chloride may cause clay to dehydrate, resulting in a decrease in permeability but an increase in porosity (Evans, 1981). Other salts may have different effects.

A rise in sea level could have two effects on land treatment areas. First, wastes could dissolve or be suspended in the nearby soil. Also, increased leachate production and migration are possible. Second, physical erosion caused by coastal wave action might result in a total washout or removal of the soil layer and the incorporated wastes.

Three primary effects on surface impoundments could be anticipated as a result of sea level rise. First, waste solution or the suspension of settled dry wastes from the bottom of the impoundment could persist after floodwaters recede. Second, physical erosion resulting from coastal wave action could cause the structural failure of the sides of the impoundment, resulting in the release of wastes. Finally, inundation could result in the migration of wastes from the facility, receding floodwaters could cause the failure of the impoundment structure because of pressure differentials and saturated soil conditions, and contact of saltwater with a clay liner could result in increased leachate movement.

A sea level rise could have three primary effects on waste piles, which employ biological decomposition in the treatment and disposal of waste. First, waste solution or suspension could occur. Second, physical erosion could occur because of high-velocity flooding, which would cause complete washout of the pile. The waste pile may remain saturated after floodwaters have receded, allowing waste to continue leaching out. Last, saturation of the pile could cause structural weakening and result in a collapse of the pile and potential washout.

Storage facilities would have the options of either developing emergency plans to remove wastes prior to flooding or incorporating structural engineering solutions. But if these measures failed, the consequences could be serious. Tanks could overflow, containers could float or spill if not properly secured, structural damage to above-ground or partially above-ground tanks could be caused by floating debris or by increased hydrostatic pressure, and saltwater could corrode tanks and containers.

Municipal waste incinerators could experience three types of damage due to a sea level rise. First, waste solution or suspension in the storage and operating components of the facility could occur. Second, structural damage, caused by hydrodynamic loads due to wave action or hydrostatic loads due to inundation could be realized, as could structural damage from floating debris in a high velocity flood. Finally, increased salt content could corrode components.

EFFECTS OF SEA LEVEL RISE ON HAZARDOUS WASTE SITES IN THE CHARLESTON AND GALVESTON STUDY AREAS

Both active and inactive hazardous waste facilities were identified and mapped for the study areas. The facilities were identified by study area zip code using the RCRA Part A data base for locating active sites and the CERCLA ERRIS file (a data base including abandoned hazardous waste facilities) for locating inactive sites. The locations of these facilities were estimated using street maps for the two study areas. The accuracy of facility location was reduced by the difficulty in plotting on the street maps, which were of very small scale. Consequently, the uncertainty in exact location must be considered, especially with regard to the analysis of facilities within the 100-year floodplain.

A characterization of hazardous wastes found at the active facilities was made using the RCRA Part A data base. Information on wastes contained at inactive sites was not available. The risk from waste types can be inferred from the presence of carcinogens and ecotoxins found at the facilities. This section discusses the changes in the vulnerability of these sites, in light of the results from Chapters 4 and 5.

Charleston, South Carolina

As a result of Charleston's industrial concentration, all of its 11 hazardous waste facilities are located in the northern and central sections of the Charleston peninsula. Figure 9-1 shows the locations of these facilities and the 10-year and 100-year floodplains. All 7 of the active sites were identified as storage and/or treatment facilities, including tanks, containers, and waste piles. The wastes associated with these facilities are listed as follows:

xylene	cadmium[a]
toluene	arsenic[a]
ethyl benzene	benzene[a]
phenol	beryllium[a]
tetrachloroethylene	chromium VI[a]
mercury	nickel[a]
cyanide	vinyl chloride[a]
nitrobenzene	dichlorobenzene
lead	iron ferro (IC) cyanide

[a]known carcinogens of greatest concern to human health

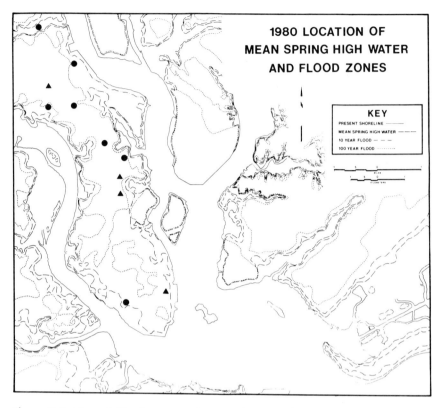

Figure 9-1. Map of Charleston study area showing the locations of hazardous waste facilities relative to the 10- and 100-year floodplains for 1980. • = active hazardous waste management facility; ▲ = inactive hazardous waste management facility.

We could not obtain information on the 4 inactive sites. However, given the stable history of industrial activity in the Charleston region, the types of hazardous waste found at the active RCRA facilities are probably a good indication of the types of waste at the inactive sites. Four active sites and one inactive site are located within the 100-year floodplain, with one active site in the 10-year floodplain.

Figures 9-2 and 9-3 show how the sea level rise scenarios would bring additional waste sites into the 10- and 100-year floodplains, respectively, for the year 2075. Under the high scenario, 5 additional hazardous waste facilities would be within the 10-year floodplain, and all but one would be within the 100-year floodplain. Table 9-1 summarizes the number of facilities (active and inactive) located within the floodplain under the high scenario for 1980, 2025, and 2075.

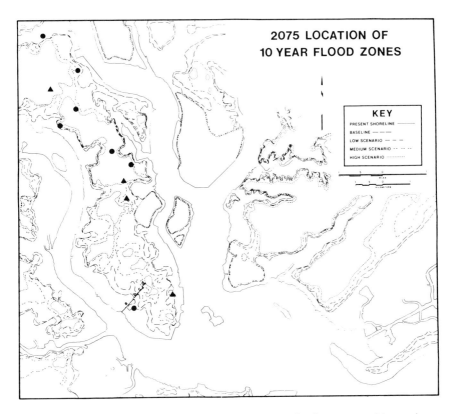

Figure 9-2. Map of Charleston study area showing the locations of hazardous waste facilities relative to the 10-year floodplain for 2075. • = active hazardous waste management facility; ▲ = inactive hazardous waste management facility.

Table 9-1. Charleston Study Area Change in Number of Hazardous Waste Facilities (HWF) in 100-Year and 10-Year Floodplains for 1980, 2025, and 2075 (based on the high sea level rise scenario)

Year	Number of HWF in 100-Year Floodplain			Number of HWF in 10-Year Floodplain		
	Active	*Inactive*	*Total*	*Active*	*Inactive*	*Total*
1980	4	1	5	1	0	1
2025	4	2	6	2	1	3
2075	7	3	10	4	2	6

Source: Data from U.S. Environmental Protection Agency, 1982, RCRA data base, Part A, and author.
Note: Out of a total of 11 hazardous waste facilities (7 active, 4 inactive)

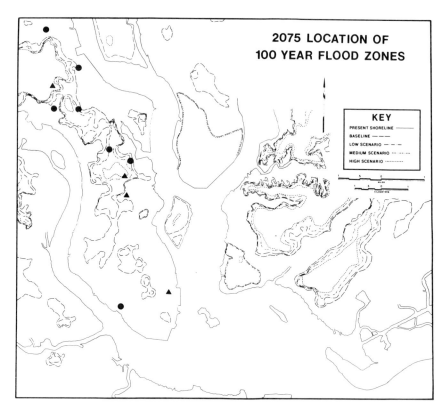

Figure 9-3. Map of Charleston study area showing the locations of hazardous waste facilities relative to the 100-year floodplain for 2075. • = active hazardous waste management facility; ▲ = inactive hazardous waste management facility.

Galveston, Texas

Thirty of the Galveston area's hazardous waste facilities are located in Texas City, and the other two sites are located in Galveston. Figures 9-4 and 9-5 show the study area's 14 active and 18 inactive hazardous waste facilities. In 1980, 10 facilities (4 active and 6 inactive) were located in the 100-year floodplain and 8 facilities (4 active and 4 inactive) were located in the 15-year floodplain. The facilities include landfills, waste piles, surface impoundments, storage tanks, incinerators, injection wells, and land treatment areas. Because most of the Galveston area is low-lying, these facilities would be vulnerable to flooding. A large number of inactive sites may be abandoned and not specifically protected against flooding, although the Texas City Levee System provides general protection.

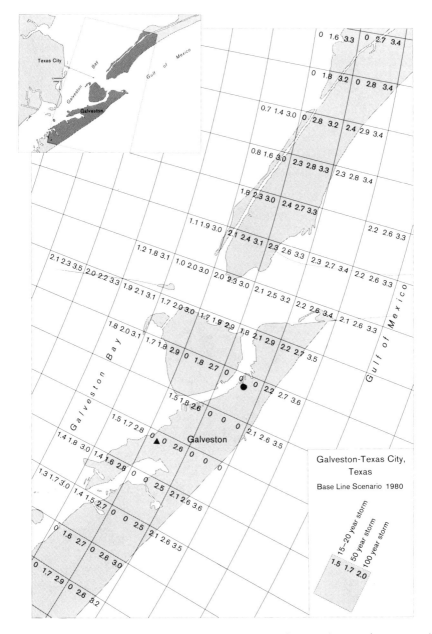

Figure 9-4. Map (Galveston section) of Galveston study area showing locations of hazardous waste facilities and 15–20-, 50-, 100-year storm surge elevations for 1980. • = active hazardous waste management facility; ▲ = inactive hazardous waste management facility.

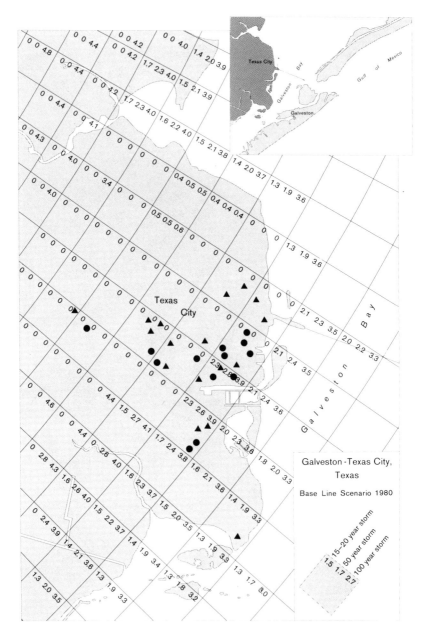

Figure 9-5. Map (Texas City) section of Galveston study area showing locations of hazardous waste facilities and 15–20-, 50-, and 100-year storm surge elevations for 1980. • = active hazardous waste management facility; ▲ = inactive hazardous waste management facility.

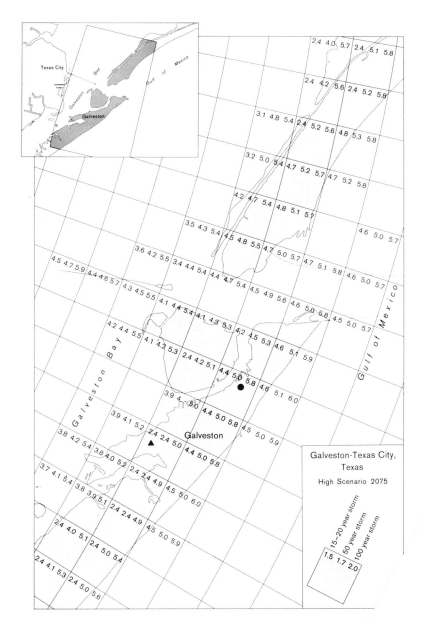

Figure 9-6. Map (Galveston section) of Galveston study area showing hazardous waste facilities and 15–20-, 50-, and 100-year (high sc surge elevations for 2075. • = active hazardous waste managem inactive hazardous waste management facility.

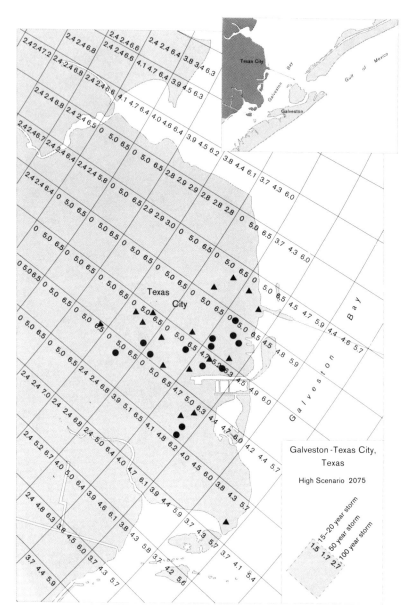

Figure 9-7. Map (Texas city section) of Galveston study area showing locations of hazardous waste facilities and 15–20-, 50-, and 100-year (high scenario) storm surge elevations for 2075. • = active hazardous waste management facility, ▲ = nactive hazardous waste management facility.

The facilities in the area containing nine known carcinogens and two ecotoxins are listed as follows:

xylene	methyl parathion[a]
toluene	lindane[a]
cyanide	benzene[b]
lead	carbon tetrachloride[b]
mercury	chromium VI[b]
napthalene	polyanuclear aromatic hydrocarbons[b]
phenol	beryllium[b]
acrylic acid	nickel[b]
nitrobenzene	cadmium[b]
iron ferro (IC) cyanide	arsenic[b]
chlorotoluene	vinyl chloride[b]

[a]ecotoxins (chemicals that are toxic to ecological systems)
[b]known carcinogens

Figures 9-6 and 9-7 show the 15-20-, 50-, and 100-year high scenario storm surge elevations and the hazardous waste sites in the Galveston study area. Thirty-two facilities (14 active, 18 inactive) would be within the 100-year floodplain by 2075, and 11 (5 active and 6 inactive) would be within the 15-year floodplain. This result assumes that the levees and seawall would not be raised, an assumption that Gibbs dismisses in Chapter 7. A summary of the number of facilities located within designated floodplains under the high scenario for 1980, 2025, and 2075 is presented in Table 9-2.

Table 9-2. Galveston Study Area Change in Number of Hazardous Waste Facilities (HWF) in 100-Year and 15-20- Year Floodplains for 1980, 2025, and 2075 (based on the high sea level rise scenario)

	Number of HWF in 100-Year Floodplain			Number of HWF in 15-20-Year Floodplain		
Year	Active	Inactive	Total	Active	Inactive	Total
1980	4	6	10	4	4	8
2025	13	18	31	4	5	9
2075	14	18	32	5	6	11

Source: Data from U.S. Environmental Protection Agency, 1982, RCRA data base, Part A, and author.
Note: Out of 32 hazardous waste facilities (14 active, 18 inactive)

Flood Control Measures for Hazardous Waste Sites in the Study Areas

If no additional floodproofing or protection measures are taken by the waste sites in the case study areas, risks from flooding, particularly in Charleston, could be substantially increased by sea level rise.

The Charleston facilities are primarily storage/treatment units; thus floodproofing options such as fencing, movable containers, elevating facilities on fill or piers, and secure storage structures could be used. Flood protection options would include constructing levees and floodwalls around facilities to cut off impeding floodwaters and digging channels to divert water around facilities. A less expensive alternative for storage facilities would be the development of emergency evacuation plans to remove all wastes before a flood.

The Galveston study area contains more diverse types of hazardous waste facilities including landfills, waste piles, land treatment areas, surface impoundments, storage tanks, injection wells, and incinerators. Thus, flood protection and floodproofing options will vary more in Galveston. However, if local governments decide to raise their seawalls and levees in anticipation of sea level rise, additional measures by the waste sites may be necessary.

CONCLUSIONS

The flood mitigation measures in place at hazardous waste facilities vary widely in their flood mitigation effectiveness. Accordingly, serious health and environmental problems may result from the flooding of some active facilities, even in the absence of sea level rise.

Although the site visits of Walash and Raasch suggest that there is a wide variation in design flood levels and in the use of freeboard, the existing flood mitigation measures have one element in common: they generally will not, in their present condition and configuration, provide protection against flooding associated with a large rise in base sea level.

The need for anticipating sea level rise and reconsidering the adequacy of existing regulations varies according to the operating status of hazardous waste sites. Owners and operators of proposed facilities should consider whether the prospect of increased flooding justifies changing the planned location. Proposed facilities may minimize flood mitigation investments in the long run by designing for the flood levels to be experienced over the project's lifetime rather than those currently to be expected. We believe, however, that existing regulations, if enforced, will protect the public from increased exposure to wastes from proposed and operating facilities. As sea level rises, flood maps should be redrawn and facilities retrofitted with additional required flood mitigation measures. The fact that FEMA has yet to complete the preparation of flood insurance risk studies for a substantial fraction of communities in the United States suggests that higher priority may have to be accorded to this function in the future.[7]

Currently operating facilities scheduled to close would not be protected by existing regulations. To address this inadequacy, closure plans required by federal and state agencies should go beyond the level of protection required of operating facilities and incorporate measures to protect these sites from the inundation, erosion, and flooding that could occur in the next century.

Currently inactive facilities, particularly those that were improperly closed or that do not have an identifiable owner or operator, may already present environmental hazards. These hazards would be aggravated by a rising sea. Identification and decontamination of these sites may pose the most troublesome of the problems discussed here.

Proven flood mitigation measures exist to address the risks that could be created by sea level rise. Environmental programs should be expanded in order to address this challenge explicitly.

NOTES

1. Active hazardous waste facilities are defined as those sites that filed a RCRA Part A application to EPA before November 19, 1980.
2. (PL 94-580).
3. Section 3004, Part 264, Subpart B.
4. 24 CFR 1910.3(c)(3).
5. The following description of flood characteristics is an adaptation from a recent EPA report (EPA, August 6, 1981) on locating hazardous waste facilities in special environmental areas. For further information, see Department of the Army, Office of the Chief of Engineers, 1972, *Flood-Proofing Regulations*, EP 1165-2-314, Washington, D.C.: U.S. Army Corps of Engineers. This study identified the characteristics of floods that are critical in determining the degree of flood damage.
6. A computer literature search of the GEOREF and Pollution Abstracts found virtually no references to the subject.
7. In April 1983 FEMA issued a request for proposals for "Identification of the Universe of Remaining Flood Insurance Studies" (RFP EMW-R-1187). Seven thousand U.S. communities that have been designated as flood prone have not had flood risk studies completed.

REFERENCES

Development Planning and Research Associates. 1982. Report to EPA/OSW.

Evans, Jeffrey C., Ronald C. Chaney, and Hsai-Yang Fang. 1981. *Influence of Pore Fluid on Clay Behavior.* Fritz Engineering Laboratory report 384.14. Bethlehem, Pa.: Lehigh University.

Federal Emergency Management Agency (FEMA). 1978. *How to Read Flood Hazard Boundary Maps.* Washington, D.C.: Dept. of Housing and Urban Development.

Hydrocomp. 1979. *Evaluation of Flood Levels for Solid Waste Disposal Areas.* EPRI research project 1260-9. Mountain View, Ca.: Hydrocomp, Inc.

Lafornara, Joseph P., Howard J. Lamp'l, Thomas Massey, and Ralph Lorenzetti. 1978. "Hazardous Materials Removal during 1977 Johnstown, Pennsylvania, Flood Emergency, Control of Hazardous Material Spills." In *Proceedings of the National Conference on Hazardous Material Spills.* Miami Beach.

Leatherman, S. P., M. S. Kearney, and B. Clow. 1983. *Assessment of Coastal Responses to Projected Sea-Level Rise: Galveston Island and Bay, Texas.* URF report TR-8301; report to ICF under contract to EPA. College Park: University of Maryland.

Research Planning Institute. 1983. *Hypothetical Shoreline Changes Associated with Various Sea-Level Rise Scenarios for the United States: Case Study—Charleston, South Carolina.* Report to ICF under contract to EPA. Columbia, S.C.: RPI.

Sorensen, R. M., R. N. Weisman, and G. P. Lennon. 1982. *Methods for Controlling the Increases in Shore Erosion/Inundation by Storm Surge, and Salinity Intrusion Caused by a Postulated Sea Level Rise.* Bethlehem, Pa: Lehigh University, Department of Civil Engineering.

U.S. Department of Army. 1972. Flood-Proofing Regulations. EP-1165-2-314. Washington, D.C.: DoA.

U.S. Environmental Protection Agency. 1980. *Evaluation of Cover Systems for Solid and Hazardous Wastes.* SW-8867. Washington, D.C.: EPA.

U.S. Environmental Protection Agency. 1980. *Standards Applicable to Owners and Operators of Hazardous Waste Treatment, Storage, and Disposal Facilities under RCRA, Subtitle C, Section 3004.* EPA background document. Washington, D.C.: EPA, pp. 106-107.

U.S. Environmental Protection Agency. 1981. "Guidance Manual for Location Standards and Special Environmental Areas." Washington, D.C.: EPA.

U.S. Environmental Protection Agency. 1982. "Land Disposal Regulations: Rules and Regulations." *CFR Federal Register* 47(143):32290-32291.

Walesh, S. G., and Raasch, G. E. 1983. *Assessing the Risks Associated with Flooding at Actual Hazardous Waste Management Facilities.* Waukesha, Wis.: Donohue and Associates.

Chapter 10

Independent Reviews

INTRODUCTION

On March 30, 1983, EPA sponsored a conference in Washington, D.C., entitled "Effects of Sea Level Rise on the Economy and Environment." Reviewers from several disciplines were asked to comment on papers that subsequently were revised into Chapters 1 to 7 of this book in light of their experience in areas that would be affected by a sea level rise. Some of these comments follow.

Sherwood Gagliano, president of Coastal Environments, Inc., reviews the case studies of Chapters 4 and 5 and the engineering techniques of Chapter 6. Included in Gagliano's remarks are suggestions for further and improved research as well as a discussion of the planning, public communication, and other efforts undertaken in Louisiana to address coastal erosion and salt intrusion.

Edward. J. Schmeltz, assistant vice president and department manager of coastal engineering for PRC-Harris, discusses some of the problems that must be considered in evaluating the engineering implications of sea level rise. Schmeltz also addresses the ability and willingness of engineers to modify their designs in anticipation of sea level rise.

Jeffrey R. Benoit of the State of Massachusetts Coastal Zone Management Program examines the possible state responses in planning for sea level rise. He discusses four options for effective state response: policy development and/or revision, regulatory reform, legal conflict resolution, and education.

Colonel Thomas H. Magness, III, formerly assistant director for civil works for environmental affairs, U.S. Army Corps of Engineers, discusses the role of the corps in maritime projects and poses a number of questions for further research. Then, Magness highlights some of the corps's research and planning efforts to develop solutions to the problems caused by sea level rise.

Charles Fraser, chairman emeritus of Sea Pines Corporation, which developed Hilton Head Island and other major coastal areas, indicates how difficult it will be to focus attention on an issue such as sea level rise, whose effects are so far in the future.

The comments presented by Lee Koppleman, executive director of the Long Island Regional Planning Board, focus on the near-term problems posed by a potential sea level rise, as well as their political implications. These implications are illustrated using examples from the Long Island coast.

COMMENTS OF SHERWOOD GAGLIANO

I am grateful for the opportunity to comment on the work centered in Chapters 4, 5, and 6, which is both well done and carefully thought out. Part of my assignment was to temper my comments with the experience gained in coastal Louisiana where, for decades, inundation has accelerated as a result of land subsidence and sea level rise. I will first discuss the two case studies—Galveston and Charleston—and then the chapter on engineering techniques. I will conclude with a discussion of the current situation in Louisiana.

Case Studies

The methods used in the two case studies were excellent. They followed a very straightforward historic approach, using what has happened in recent decades as a basis on which to project future conditions. They combined this historic approach with a process form approach, where some adjustments were made based on current knowledge of process-response models in the coastal zone. I particularly liked the way the studies used the computer to manipulate large quantities of data and made adjustments based on both judgment and experience.

The results and the projections of both studies are very acceptable. If anything, they are conservative but still provide a basis for making decisions across a wide range of scenarios. Of course, our ability to project shoreline changes can be improved as the method is applied to other areas.

First, as Kana et al. and Leatherman realize, the data base could, as always, be improved. The quality of information available in map form — topographical maps particularly — is insufficient for many coastal areas. In far too many cases, the finest resolution available in contour maps is 5 ft. Many need updating and some of the maps that have been updated are inaccurate.

Measures are necessary to improve the quality of elevation data in all coastal areas. For example, survey lines that have been carried in from stable areas have often not been resurveyed in several years. Assuming that coastal areas are stable is a mistake. In Louisiana we have come to realize that the land surface is like the surface of a waterbed. Some areas are sinking, and others are rising. The loci of these uplifts and subsidences are changing over time, and the rates of both uplift and subsidence are significant in reference to the anticipated rates of sea level rise.

Another factor that needs more consideration is the possible change of coastal river regimes that may accompany sea level shifts. Long-term geographical and archaeological models suggest that during the transition from a standstill to rising conditions and vice versa, changes in the total discharge of rivers, the magnitude of the floods, and the amount of transported sediment can be expected. These changes would result in additional impacts in coastal areas.

The most important conclusion of the case studies is that by using readily available information, state-of-the-art modeling techniques, and an understanding of coastal processes, it is possible to develop in a relatively short time at least a first-cut prediction model for any U.S. coastal area. It would be a mistake to wait until modeling techniques are perfected before we start applying this technique and incorporating its results into the planning cycle.

The results of the case studies for the base conditions and the three scenarios are not surprising. As I understood the two case studies, there are no catastrophic results expected under the continuing base conditions, the low scenario, or the medium scenario through 2025. But if the high scenario should start to unfold, some serious problems can be anticipated as early as 2025 in both of the case study areas. By 2075 there would be serious problems in low-lying coastal areas under any circumstance, with the problems being greatest under the high scenario. Low-lying coastal areas like Louisiana will suffer much more severely from the effects predicted than areas with higher relief.

Both pilot studies and the engineering study considered three types of impact: the changes in shoreline position due to erosion and inundation, the changes in frequency and depth of flooding due to storm surges, and saltwater intrusion in potable aquifers in the coastal zone. Any of the scenarios would decrease the high productivity and the diversity of fish and wildlife in wetlands, estuaries, and such fragile habitats as dunefields. Geological studies reveal that during periods of sea level rise, landforms become simplified and fewer in number, resulting in less diversity in the kinds of habitat that appeared in the coastal zone.

How sea level rise would affect navigation also warrants careful attention. Flood protection requirements and the potentially increased need for direct surface water to offset effects of saltwater intrusion into aquifers will clearly necessitate many more water control structures in coastal navigation channels. The plumbing in our coastal zones could be very intricate in 50 or 100 years. As we all know, these are very costly structures and represent a major factor in looking at the economic impacts of sea level rise.

We should also think about how sea level rise will affect historic and archaeological sites and recreation areas. South Louisiana has already experienced a dramatic loss of coastal archaeological sites during the last 20 years as a result of shoreline erosion and inundation. Higher priority should be placed on inventories of these kinds of resources.

Engineering Techniques

I find Chapter 6 to be both appropriate and straightforward. The techniques described by Sorensen et al. have been tested and proven in the field. The chapter also presents important and representative cost estimates. The authors present an entire arsenal of hard and soft structural techniques for fighting shoreline erosion. If these sea level rise scenarios really unfold, there will be a need to use every weapon in that arsenal.

Nonstructural alternatives have not really been treated in depth because they were beyond the scope of the work. However, the topic warrants another symposium of this type, involving zoning, floodproofing, abandonment and relocation, and temporary evacuation during storms.

The engineering approaches reveal a strong bias toward working on shorelines with sandy beaches and barrier islands. Relatively little coastal engineering attention has been directed toward muddy and carbonate coasts. For example, how do we manage coastal wetlands in the face of a rising sea level? Our experience in Louisiana has revealed that they are not self-maintaining.

Shoreline retreat measurement alone cannot be used as a basis for predicting coastal wetland deterioration and loss. It is necessary to map the vegetation habitats in great detail and apply a predictive model that uses environmental succession to spot areas where the wetlands are under stress and starting to break up into open water.

We should look at what the Dutch have been doing since about 1200 A.D. to protect themselves against the kinds of problems that are discussed in this volume. This includes the construction of dikes and drainage systems, intricate navigation systems that involve locks, and water control structures that totally manage surface water movement for many purposes. Many situations will require developing polders comparable to those in the Netherlands. These will be very costly, but if we examine the cost-benefit relationship of the Dutch approach, they will prove to be favorable. The Netherlands, which has one of the highest population densities on earth, has consistently enjoyed a good economy and has thrived socially and culturally.

Subsidence in Louisiana

About 15 years ago, we realized that the 5,000 year trend of progradation, resulting from sediment deposited in the Gulf of Mexico by the Mississippi River along the coast of Louisiana, has ended. In its place, we are now experiencing massive shoreline erosion and land loss. While much of the loss is man-induced, subsidence or apparent sea level rise has also been a contributing factor. We have now documented a loss of about one million acres since the turn of the century and a current rate of loss of 129 sq km (50 sq mi) a year. Measurement techniques similar to those presented in Chapters 4 and 5 have been used in determining this rate; we are only now getting to the point of using these findings to project the future shoreline.

Louisiana is a good model for what other low-lying areas can expect as a result of sea level rise. In the last 40 years there has been not only a loss of wetlands, but also a sinking of ridgelands (natural levees). These are the foundations, the high ground upon which most of the development in southern Louisiana is located. Thus, Louisiana faces a very serious problem. When one notes that the highest natural elevation in New Orleans is only 13 ft above sea level and that sinking rates of at least 3 ft per century are occurring in some parts of coastal Louisiana, one begins to realize the enormity of the problem. The barrier islands are breaking up in the washover mode discussed in Chapters 4 and 5. Their loss will result in the shorelines retreating 40 to 48 km (25-30 mi) in some instances and moving into close proximity of some of our urban areas. Also, the Louisiana coast produces about one-fourth of the nation's annual fish harvest. Most of the utilized species depend on the estuaries that are disappearing so rapidly.

Louisiana also provides valuable lessons in responding to the problems associated with sea level change. It took a number of years for the scientific community to recognize and define problems related to sea level effects and even longer to convey these findings in a meaningful way to the public and decision makers. In Louisiana, members of the scientific community have learned that there is a delicate balance between being a harbinger of catastrophe and a rational scientist. It is important to communicate facts and information to decision makers without causing panic. We have had the most success in relating impacts to specific places. For example, when land loss rates of 129 sq km (50 sq mi) per year, in the Louisiana coastal zone were first disclosed, it was newsworthy and of concern. However, it was only after disclosure that a given coastal parish would last only 50 years before it eroded into the sea that the state legislators and the governor enacted a program for coastal erosion protection and shoreline restoration.

In Louisiana we have embarked on a course of responsible planning to deal with these natural hazards. For example, the state has a federally approved coastal management program that is responsible for permitting coastal activities as well as coordinating planning and management. There is reasonably good cooperation between federal, state, and local agencies in dealing with these problems. Some imaginative plans for diverting freshwater, restoring barrier islands, and managing marshes have been developed, and pilot projects are being implemented. A realistic multi-use management approach that recognizes the need for a balance between carefully planned and executed development and conservation management is being taken.

A responsible effort is also emerging in the private sector. A number of developers and landowners in coastal Louisiana are incorporating environmental design features in their new projects. These will not only protect the development areas from future storm surges and flooding but will also ensure that the projects will be compatible with conservation management objectives of adjacent estuarine and marsh areas. Owners of large tracts of wetlands are developing marsh management plans to sustain their properties.

While the potential problems are very large and challenging, the resources and opportunities of coastal areas will continue to attract large populations. It is unlikely that coastal areas will be abandoned because of the threat of natural hazards. In some instances we may be forced to flee to higher ground, but where there are large capital investments and compelling reasons for people to be in the coastal zone, the areas will be maintained.

COMMENTS OF EDWARD J. SCHMELTZ

To a design engineer responsible for the development of various coastal and waterfront structures, the prospect of a significant rise in sea level over the next 50-100 years is a major concern. The implications of such a change in water levels on coastal regions and facilities are numerous.

These remarks will address some of the primary effects of a rise in sea level on factors relevant to the design of coastal structures. A discussion is also presented on the ability and willingness of design professionals to implement changes in design in anticipation of sea level rise.

The Engineering Implications

Significant changes in sea level can have profound repercussions on the planning and design of coastal facilities. Although the factors identified below are not an exhaustive list, they indicate the types of problems that must be considered in evaluating the engineering implications of sea level rise. Prevention, mitigation, and response techniques for a variety of coastal structures have been addressed by Sorensen et al. in Chapter 6.

A critical factor in the analysis of coastal structures is the stillwater level utilized to evaluate design conditions. The magnitude of elevation of the water surface during a storm has a direct bearing on such factors as inundation of low-lying areas, the point of impact of wave action on structures, and the level of wave attack on beach/dune systems.

Less obvious secondary effects may also occur. Examples include increased breaking wave heights, variations in tidal amplitudes and phasing, modifications in storm surge elevations due to increased water depths at a given coastal location, and the effects on water tables.

If ocean temperatures increase as indicated in Chapter 1, the impact on hurricane generation could also be critical. Such storms forming in more northerly latitudes, combined with possible changes in storm tracks, could result in significant modifications to storm surge patterns and wave heights; knowledge of these phenomena is used in the design of various waterfront facilities. In some cases, these consequences could be more critical than the direct effects of sea level rise.

Two basic issues must be addressed, however, in order to assess the response of the coastal engineering community to sea level rise: technological capability and the motivation to institute changes.

Technological Capability

From the standpoint of coastal structures, the technology required to deal with changes in sea level is not new. The capability exists within the engineering community to deal with variations in water level and the attendant changes in design conditions.

The magnitude of sea level rise, as well as the rate of change is, however, a critical factor. From an engineering standpoint, it is not adequate to predict increases ranging from 2-10 ft. Rather, a more definitive assessment is necessary, keeping in mind that the application has far more practical implications than scientific interest.

A 2 ft rise in the next century would, of course, have far less impact on design than a 10 ft rise. In many areas and for a variety of structure types, a 2 ft rise would probably be of little consequence. For example, some regions of the U.S. coastline are subject to high storm surges. Design stillwater levels on the order of 4.5-6.0 m (15-20 feet) for a 100-year storm are common. A 2 ft increase is not only overwhelmed by other factors but also falls within the combined accuracy of numerical methods of water level and wave height prediction commonly used to establish design conditions. In fact, it is probably safe to say that most structures have sufficient safety margins to withstand a 2 ft sea level rise and resulting increases in wave height. As an example, at a site where waves are breaking, controlled heights would increase approximately 0.5 m (1.6 ft) for a 2 ft increase in stillwater level.

On the other hand, a change of sea level on the order of 10 ft is far more consequential. It is unlikely that most coastal structures could withstand this magnitude of change and still serve their intended function. The effects of inundation are obvious and were discussed earlier. In some areas, a 3 m (10 ft) rise exceeds the total 100 year flood elevations currently accepted. In Long Island Sound, for example, 100 year flood elevations range from 2.4-4.0 m (8-13 ft).

The magnitude of the change in sea level is, therefore, critical in assessing the impact on coastal facilities. Clearly, there are cases where a 2 ft rise is important; in other cases, the change in sea level would have to approach 3 m (10 ft) before significant effects occur. The consequences must be evaluated on a site-specific basis. Reevaluation of existing structures will be necessary to ensure their stability and adequacy to serve their intended function as water levels rise. For new structures, it is a relatively simple proposition to increase design water levels and evaluate other design loading conditions on this basis.

Motivation to Institute Changes

Although the technology exists to address significant changes in sea level, designers of coastal facilities and structures must be convinced of both the probability and the magnitude of future increases in sea level. Approximations and conjecture on the part of scientists may not induce the engineering community to institute necessary design changes.

The credibility of the analysis of possible sea level rise is probably a more critical issue than the actual numerical results of that analysis. In order to understand the engineering profession's attitudes on the subject, one must understand two factors: the engineer's position in the decision-making hierarchy of a project and the consequences of an individual engineer's belief in a significant sea level rise.

An engineer's relationship with the client/owner is a special one. The engineer provides the technological interface between the client/owner and the scientific community. While it is his or her professional responsibility to develop technically adequate designs, he or she must also be sensitive to the project's other needs, such as scheduling and costs. His or her primary function is to translate scientific facts into the hard realities of, for example, a structure. At the same time, the engineer must maintain a proper perspective on such factors as construction cost and start-up requirements. Although the engineer develops and evaluates alternatives, detailed designs, cost estimates, and specifications, the client holds the ultimate decision-making authority. Recommendations made by the engineer are just that and must ultimately be justified to the satisfaction of the owner. This is particularly true of matters that tend to increase project costs.

As Chapter 1 indicates, coastal facilities tend to be evaluated in terms of their economic life, typically taken as 30 or 40 years, even though their true useful life may be substantially greater. Convincing a client to expend additional monies to accommodate a sea level rise that "may" occur over the next century would at best be difficult. Basing this recommendation on an analysis that results in a range of 0.6-3 m (2-10 ft) stemming from input conditions that could be viewed as conjecture, would be essentially indefensible from the client's standpoint.

Clearly, a long-term view would be required of both the client and the engineer to modify design conditions to accommodate a substantial sea level rise. Unfortunately, this type of approach would not be in keeping with the economic realities of the times, unless there was firm belief in the probability of sea level rise.

Presupposing that an engineer develops a firm belief in a substantial sea level rise, it is interesting to consider the consequences of this belief. The engineer is placed in the position of convincing not only a client but, in many cases, international lending institutions such as the World Bank. There will also be, in many cases, a considerable impact on a project's initital costs. These costs are critical in the economic decisions that determine a project's viability.

If the engineer's recommendation is not founded on a credible basis that is generally accepted by his or her peers, this task will be difficult. As a simple example of the magnitude of the impact on costs, consider the rubble mound breakwater shown in Figure 10-1. With a crest elevation of 10 ft, side slopes of 1:2H, and located in water depths of 10 ft, it is essentially typical of jetties and groins at many coastal locations. The

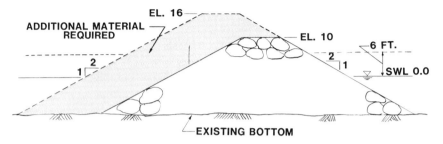

Figure 10-1. Effect of 6 ft (1.8 m) increase in height on breakwater cross-section.

dashed lines in the figure indicate the shape of the structure for a 6 ft increase in stillwater level, excluding the effects of increased wave heights and run-up. For this simple case, the volume of the structure increases by more than 60 percent for a 30 percent increase in structure height, as indicated by the shaded area in the figure. The cost impact is obvious.

There are other practical effects related to instituting changes in structural design to accommodate future sea level rise. For example, port deck elevations in the future would be higher than those currently in use. If these requirements were incorporated in current designs, new sections of wharfage could be up to 3 m (10 ft) higher than adjacent, existing sections. The movement of cargo along the structure would be complicated by this "step." Convincing the owners of a port where all existing berths are a constant elevation above mean sea level to raise new facilities 10 ft will be extremely difficult without a more credible justification.

Summary and Conclusions

The response of the coastal engineering community to potentially significant changes in sea level is not primarily a technological issue. Rather, it is a question of credibility.

The technology exists for the professional to respond adequately to the phenomenon. Design engineers must be p1 ovided with reasonable estimates of the magnitude of the problem. The support for these estimates must be sufficiently detailed to convince the engineer to recommend design changes that may result in significant increases in project costs. Clients and lending institutions who provide the funding for major marine projects must also be convinced of their credibility.

The scientists involved with the evaluation of the phenomenon must remember that there is far more than scientific interest at issue; design engineers must live in a world of cold, hard realities. Project decisions are made on the basis of technical facts and economical considerations. Conjecture and hypothesis are not sufficient, and additional research will be necessary before engineering decisions are likely to incorporate sea level rise.

COMMENTS OF JEFFREY R. BENOIT

The results from Chapter 3 indicate that by the year 2100, the level of the sea could rise over 3 m (10 ft). A rise of this magnitude would result in direct changes in biological, physical, economic, and social systems within the coastal zone. The responsibility for dealing with these changes will ultimately have to be shared by both the private (landowners) and the public (government) sectors.

Coastal states will be faced with many controversial and expensive decisions if the initial results of the EPA study prove true. Individuals responsible for making those decisions should begin to consider what components of a response plan are most suitable for their own state. These comments are intended to provide some preliminary thoughts on this subject and in doing so present a framework of options from which state governments could formulate a comprehensive plan to respond to sea level rise.

Before discussing potential response options, however, a brief look at Massachusetts will reveal several areas of concern that may also be faced by other states.

First, Massachusetts has had a long history of coastal storm damage. The most destructive coastal storm of recent times was the great blizzard of 1978. This northeaster battered the Massachusetts coast for three days. Over 330 homes were substantially damaged or destroyed, and financial losses along the coast were in excess of $180 million. Historically, attempts to protect coastal property relied on the construction of massive structures for shore protection and flood control. Because these types of protective structure were and continue to be designed only for existing sea level conditions, their effectiveness against increased sea levels will be progressively diminished.

Second, the protection of coastal wetland areas has also been of great importance to the citizens of Massachusetts. Strict laws and regulations protect valuable resource areas such as beaches, dunes, salt marshes, and barrier beaches. Changes in sea level will result in rapid spatial shifts of the boundaries of these resources. A revised and expanded regulatory approach will be necessary to ensure their long-term protection.

Finally, most coastal communities in Massachusetts accommodate recreational boating and support commercial fisheries. Consideration must be given to the impact that rising sea level will have on the usefulness of piers, docks, and other support facilities. Large urban cities along the coast, like Boston, are also major shipping ports and represent regional commercial and financial centers. As a result, many of these cities contain sizable airports, such as Logan International Airport, located in the middle of Boston Harbor. Logan Airport and much of downtown Boston are reclaimed tidelands, resulting from filling activities that began in the late 1600s. Not surprisingly, Boston's waterfront has little existing freeboard with which to accommodate a rising sea level. Thus, the modification and upgrading of shoreline protection structures will have to be appraised.

Response Options

A variety of options exist through which state governments could effectively respond to increasing sea level. Although all the options may not be suitable for application by each state, collectively they provide a framework for states to choose from. This framework can be divided into four categories: policy development and/or revision, regulatory reform, legal conflict resolution, and education. A brief discussion of each option follows.

If state agencies are to be effective in their long-term planning efforts to mitigate projected sea level rise impacts, new policies must be developed and existing policies revised. A shift in state policies can be achieved in a variety of ways such as gubernatorial executive orders, individual agency response and, interagency coordination. Policy changes can deal with such issues as the elimination of state funding or incentives for new development within areas to be affected by sea level rise; the revision of post-storm recovery policies (evaluating existing rebuilding practices); and the development of criteria to identify, evaluate, and assign priorities to areas for receipt of public funds for construction or modification of shoreline protection structures.

The second category of options consists of changing existing regulatory requirements. A variety of changes could be implemented including the mandatory review and updating of existing coastal wetland inventory maps; coastal resource protection legislation and regulation based on dynamic boundaries as opposed to static mapping products; the revision of state building codes to recognize a progressive landward shift of hazard zones; and the inclusion of sea level rise projections in projects subject to review under state environmental policy acts.

As shoreline positions shift in response to a rising sea level, conflicts over ownership and property rights will arise. Many of these potential conflicts could be avoided prior to their creation if state tidelands statutes were clarified. Specific attention should be directed to the burden of responsibility for impacts on adjacent property resulting from the use of shoreline stabilization structures and to the private right to erect shoreline stabilization structures in contrast to the public right to use the beach.

The fourth, and final, category is education. Introducing a new idea to any audience must be accompanied by an informative explanation. This especially holds true when an idea presents a radical change from accepted concepts. The general public, private investors, engineers, architects, and public officials all must be made aware of the problems associated with sea level rise projections. Accepting the serious nature of the problems is essential if long-term planning measures are to be successfully implemented.

Conclusion

If the high scenario predicting a 4 m (12 ft) sea level rise by 2100 is true, coastal states must begin formulating response plans immediately, but before any action will be taken by legislatures or state agencies, there must exist a strong scientific information base. Without defensible data, it will be extremely difficult to make any substantial changes in state government or to convince the general public that a real problem exists. The extent to which future sea level will change and just what the consequences of the change really mean must be more accurately determined. However, the value of EPA's work in this area is that it will force state agencies, planners, and the public to ask the difficult questions posed by this project. Then we can begin to elicit answers from decision makers via increased research and data gathering efforts.

COMMENTS OF THOMAS H. MAGNESS III

When Chicken Little warns us that "the sky is falling," we in the Corps of Engineers do not worry much. We figure that must be NOAA's or NASA's turf, if you will. But when we get reports that the sea level is rising, we see Chicken Little addressing a water resource problem and take notice.

The Corps of Engineers has had a long and "meaningful" relationship with the sea. The bulk of our maritime involvement has been in the areas of navigation (at ports and inlets), flood control (including floods by hurricane surges), and shore protection (which sometimes involves beach restoration). We plan over the meso-term: longer than a storm event but shorter than a geologic period. Most of our projects in these areas are designed for a useful, functional life of 50-100 years. This is about as far into the future as we can predict, with some certainty, the complex combinations of physical, economic, environmental, and social interactions

Note: This paper does not necessarily represent the official position or policy of the Department of the Army.

that will have an impact on a given location. Within that time frame, we are interested in emergency actions, planning, and research on any natural phenomena (such as changes in sea level) that would have an impact on the effectiveness of our ongoing, planned, or existing projects and on the Corps's ability, in general, to carry out its missions.

In order for any of us to examine a change in the sea level relative to the land mass, we obviously must identify each of the processes that could cause such a change, quantify each process, and then re-examine the problem as a whole once the relative importance of each process has been made as clear as possible.

The work presented in this book has certainly accomplished part of this requirement. There is now a more complete list of the individual processes that affect the relative positions of land and sea. We now have a good idea of the present level of knowledge of each of these processes. There is now a first cut at trying to quantify one of these processes, namely, the global warming that results from the build-up of CO_2 in the atmosphere. Finally, there is a feeling (though perhaps not yet a consensus) that the relative importance of that process is increasing with respect to other processes.

There are still, however, a number of questions that need to be addressed. Is the list of all the processes involved complete? While it appears to be complete, if we really want to be sure, we need more forums such as this one where all disciplines can interact. What efforts are underway to quantify other processes that affect global warming either positively or negatively? An increased level of effort must be put into research on other, potentially significant phenomena. At this stage, most such studies must still be considered largely speculative. What is the relative accuracy of the quantification of the various other processes? It would do little good to have greater accuracy in quantifying a relatively minor process. This question does not seem to have been addressed yet. What is the relative importance of each of the processes in the "big picture"? How much of the rise in sea level is due to global warming, and how much might be due to something else? There is, as yet, no generally accepted apportionment. How accurate are the measurements of sea level? This has been a continuing problem to many of us, since most tide gauges have historically been installed in bays and sounds. Further, there is also some evidence suggesting that sea level might be responding to geologic-scale oscillations, rather than rising generally and continuously. How do changes in climate and sea level affect the processes that cause beach erosion? The forces required to move beach sand are generated by storms, not by long-term sea level rise. Long-term sea level rise primarily affects the elevation at which the erosive forces act. Long-term climatic changes (such as an increase or decrease in the number of severe storms) would have a far greater effect on local erosion rates. Has anyone yet studied scenarios for estuaries, coastal marshes, and associated freshwater wetlands in response to various sea level rise scenarios?

The Corps' basic missions, of course, are navigation, flood control, and shore protection; these missions dictate the focus and extent of our consideration of sea level rise. These basic missions are, likewise, supported by a considerable amount of applied research by Corps laboratories. The Coastal Engineering Research Center, the Waterways Experiment Station, and the Cold Regions Research and Engineering Laboratory have been working to apply the results of basic research, along with extensive field and laboratory data, to the behavior of inlets and channels, beaches and dunes, and estuaries and wetlands. We must better understand the natural behavior of these features before we can accurately predict their behavior after the installation of man-made projects.

The Corps has already been studying the effects of sea level rise. In 1980 the Coastal Engineering Research Center began a program to increase and refine the basic knowledge of barrier island processes. One of the major ongoing substudies of this program examines sea level elevation changes as they affect barrier island processes, in an attempt to predict future changes.

It is also obvious that beaches have multiple uses, and to plan and design a beach erosion control or hurricane protection project properly, we must consider the local short- and long-term sea level rise (or fall). This book does not address sea level fall; however, it is the Corps's experience that sea level change is very much a site-specific phenomenon. There are many places where a perceived sea level fall has been observed (Astoria, Oregon, and Juneau, Alaska, are well-known examples). At other places, like Texas City, Texas, and Long Beach, California, subsidence has caused a perceived sea level rise. Because the causes of such changes are not well understood, either qualitatively or quantitatively, generalizations about the effect of global sea level rise, from our experience, would be tenuous at best. Yet, we must and do consider sea level changes, however difficult they are to quantify.

The Corps's New Orleans District is currently involved in a study of the Louisiana coastal area, where land subsidence and global sea level rise are but two of several phenomena simultaneously affecting change. In areas such as Louisiana where the relative local change is much greater than the average "global" sea level rise, the Corps was asked to develop solutions to the resultant local and regional problems. While the results of research into global sea level rise will be useful in our planning and design efforts in Louisiana, solutions to these problems should stand us in good stead if we must later address a more general rise.

The Corps's Los Angeles district is currently involved in a regional study of the southern California coast to identify and to quantify all the processes affecting that coast. These regional study results will also be available for future, more localized or broader-scale planning investigations, with research on global sea level rise, or fall, fed into the process models developed in that regional study effort.

The complex hydraulics at inlets, particularly those where there is a navigation interest, has caused the Corps to predict the effects of sea level rise on the basis of the processes occurring there. Not only are navigable waterways affected but also the adjacent beaches, shoals, dunes, and lagoons. We must understand all the processes interacting there and the effects of our navigation works on those processes. Thus, a better capability to predict future sea level conditions is crucial to the adequate planning and design of navigation channels and harbor improvements.

Although our concerns for navigation, flood control, and shore protection form the bulk of the Corps's interest in the coastal zone, we do have some other interests as well. The Corps permit programs, authorized by Section 10 of the 1899 River and Harbor Act and Section 404 of the Clean Water Act, are important ones. Those programs require the Corps to investigate and then certify that an applicant's proposed construction, dredging, or land reclamation in waterways, wetlands, or along the coast would not seriously degrade or otherwise harm water quality, the environment, the property of others, or the public welfare (including navigational interests).

Another all-pervasive concern of the Corps is the individual states' approved coastal zone management plans. We in the Corps do not manage any of the coastal zones per se, but we do conduct our business in concert with state plans and we must certify that our works are not inconsistent with those approved state plans.

We have also been asked by EPA to help in the clean-up of many of the nation's hazardous waste disposal sites. As we have heard, many of the "Superfund" sites are in the coastal zone where a change in sea level would certainly affect where, when, and how we design and accomplish a particular clean-up.

Because of the way Corps studies are authorized and funded, we can say with some certainty that our planning and design process is of sufficient length to allow for the incorporation of any potential general sea level rise in a rational and orderly fashion. And in fact we have already begun.

We plan to share our research results and the findings of site-specific planning studies with EPA and others. Likewise, we hope that any research results that others develop will be complementary to our own research and planning efforts.

As you may surmise from these remarks, the Corps of Engineers has a very definite interest in sea level rise. We want to and we must better understand all the complex interactions of the phenomena that cause the sea level to rise or fall. Such an understanding will allow us to predict with greater confidence into the next 100 years and therefore to better plan and implement needed water resources projects in the coastal zone. EPA's start on this program deserves the support and encouragement of us all. To the extent that our resources and mandates permit, the Corps can be counted on to cooperate—and to do so fully and actively—for it would be folly not to believe that Chicken Little only has to be right once.

COMMENTS OF CHARLES E. FRASER

The world is, unfortunately, filled with examples of tardy responses by citizens affected by significant adverse changes in their immediate work and living environments.

The coastal shoreline of America was of little value, except for occasional ports, in the era when agriculture dominated our economy. In 1900 the 48 km (30 mi) of coastline in South Carolina known today as Myrtle Beach were regarded as worthless. Owned by the Myrtle Beach Farms Company, the sandy land near the ocean was not good for either growing tomatoes or growing good pine trees. It was valued at about $2 an acre, but pine land 10 mi inland was valued at about $10 an acre.

As recently as 1950 on Kiawah and Hilton Head Islands—neither of which had bridge connections—the oceanfront acreage was valued only for the value of the standing timber of those islands—about $100 an acre. Today, 34 years later, as a result of superb land use planning and tight controls in the "golf plantations," oceanfront property sells at $400,000 per half-acre lot, despite threats of hurricanes, rising waters, and high taxes.

For 20 years Congress rejected the idea of buying Cumberland Island for $500 to $1,000 an acre, despite the Mellon Foundation's *Vanishing Shoreline* book of 1954. The U.S. government is assembling the final pieces and is now buying it at $15,000 or more an acre.

Because people freed from the farm and factory burdens of 70 hour work weeks began to visit the seashore, beach railroads became very profitable investments. One could run a railroad on piling and crossties over marshes, out to a beach, load up all the passenger cars available in Atlanta or Philadelphia, and run an excursion express train to the closest beach. The end of the railroad at the station closest to the beach produced a 1920s boom in land values.

A new fad developed after World War II. Tanned skin became a symbol of affluence and leisure. People began to expose their skin to the sun, and the surge of people to the shore increased about fivefold. An interesting bit of trivia is that within the first 5 years of the redevelopment of the ancient Roman bikini in 1947, the summer crowds on the French Riviera increased tenfold.

Predicting trends in the economy of the coast is very difficult. One really does not know what land prices are going to be. They certainly will not plunge downward as a result of this conference. Even though I have been studying the history of coastal values for many years, I have managed to make enough miscalculations to develop a sense of humility. In 1968 I calculated that 10-15 percent a year price increases were at an end at Sea Island, Sea Pines, Hilton Head, and Myrtle Beach and that thereafter they would appreciate at only 5 percent per year, so I sold out. With great wisdom I bought 6 miles of beach in Puerto Rico in 1969. Since I made that then seemingly brilliant decision, the land that I left behind at Hilton Head Island has appreciated 1,000 percent and the land in Puerto Rico has depreciated 80 percent. But I am learning.

As for sea level rise and fall, which are so characteristic of our earth, the concern about another 6 in or even 2 ft rise over the next 20 or 100 years is less today on islands than the current uncertainty about the height of the next hurricane surge tide and the speed of evacuation. Houses at Hilton Head Island are already being built 8 ft above the highest known hurricane wave of the last 300 years. A new computer model—whose authors admit that it contains enormous numbers of gaps in the data—predicts that the next hurricane waves will exceed that record by 10 ft. It's unlikely that a vast program of dealing with an issue that may possibly surface 20-30 years from now is going to attract the attention of a community that is not even willing to deal with the immediate present's problem of over-building, which threatens safe hurricane evacuation.

For example, we tried to get the Beaufort County Council for the past 4 years to react to the explosive growth on certain parts of Hilton Head Island. We have both the best and the worst planning in America on Hilton Head Island. Eighty percent is superbly planned, 5 percent is being raped, and a question mark hangs over 15 percent. For 4 years the council has been debating a land-use ordinance that was unanimously recommended by organizations ranging from the Sierra Club to the board of directors of the Chamber of Commerce Hilton Head. So a county government body that cannot even bear limited controls is apt to have difficulty dealing with problems as complex, as subtle, and as far beyond the next five elections as sea level rise. We will vote in a new city government for Hilton Head on August 2, 1983, and tackle such problems locally.*

*The results of this election instituted a town government for Hilton Head Island.

Similarly, in New York City, you would think all the city agencies would have leaped to support the Rouse Company in the South Street Seaport. It was proposing to carry forward the South Street Seaport development after the brilliant success of the Quincy Market program in Boston and Harbor Place in Baltimore. Yet it took the Rouse Company 10 agonizing years just to get the state and city agencies to agree on a few very simple things for the South Street Southport Connection.

At Hilton Head Island, Sea Island, Georgia, or any of the low-density coastal areas with single-family houses, one might suggest that the owners give up their cherished guarded gates and private drives to get public funds for a seawall to guard against the possible rise of the ocean 10 or 20 years from now. But since many of them are retired, they would say that the sea level would not rise in their lifetimes, so the "Feds" and everybody else should leave them alone.

Similarly, try proposing to the Historic Commission of Charleston that some agency build a 3 m (10 ft) wall around some historic sections of the city. For 22 years a local DAR group resisted restoring the old Exchange Building at the foot of Broad Street, once used to hang pirates. The 22 women on the DAR Board did not want a lot of blacks inside the building 20 years ago and blocked federal funds for years. Imagine the local response to a $50 million bond issue for seawall.

The capacity of institutions to resist complex social or civil engineering projects and the potential for protracted debate over who should pay are almost terrifying whenever action is really urgent. I believe the sea level rise must become more visible before people will take the problem seriously. And I do not think scientific papers are the answer. It will probably take three good hurricanes, which will take us about 50 years. Then, in the rebuilding process, we will address sea level rise. In the meantime, since it will probably take several years to prepare the designs for the creative work you are doing on sea level rise, let's go forward and try to compress a 50-year cycle down to a more meaningful 25-year cycle. At least we will thereby save 25 years, which is not perfect but is better than not saving any time in our inevitably slow reaction to the world around us.

Is the sea level actually rising today? Despite the $200,000 study reflected at EPA's sea level rise conference, the issue is very much in scientific debate.

COMMENTS OF LEE KOPPLEMAN

After reviewing the previous chapters, I recalled the story of a community suffering a devastating flood. The minister was seen hanging out of the second-floor window of the church. Someone in a passing boat yelled, "Reverend, jump in—we'll save you"! He said, "Go on, fear not, the Lord will save me. Save others." After turning down similar offers and climbing higher and higher in the church, the minister found himself on the top of the steeple, clinging to the cruciform. A rescuer in a helicopter came by, telling him to grab a rope. Again he said, "Fear not, I trust in the Lord." In the next scene the minister was in Heaven. The Almighty welcomed him, and the Reverend said, "You know, I appreciate being here in Heaven, but I'm a bit frustrated. All my life I've served You faithfully, but in my time of need You did not respond." And God replied, "You, frustrated? First I sent one boat, then I sent another boat, . . . and then I sent a helicopter"!

I believe that the first boat passed us planners by. In fact, I'm afraid the second may have passed us too. In a free society, we only seem to respond politically when faced with crises. Dr. Sherwood Gagliano says we should neither resort to hysteria, nor alarm the public. I thoroughly agree. Government should lead, guide, and comfort its citizens. However, this responsibility does not call for ignoring a crisis when it appears.

Some chapters in this book indicate that perhaps we have no near-term problem. Others mention that sea level rise may increase a few inches or a few feet. When reading them, I was beginning to feel very comfortable, thinking we do not have a problem. But the more I thought about it, the more I realized this interpretation is incorrect and that complacency is the wrong direction. The great value of the papers presented here is that they illuminate the problem.

For example, years ago when I served on the National Coastal Zone Advisory Committee, I remember Dr. Lyle St. Amont saying that there was no conflict between Outer Continental Shelf activity and its impact on Louisiana's wetlands. After all, Louisiana has 3 million acres of wetlands. But with the current annual loss of 13,050 hectares (32,000 acres) of wetlands, in 100 years—just about coincidental with the time horizon of this study—Louisiana may bid its wetlands goodbye.

Thus, the key question is: what are the implications of the work that the very talented people participating in this study have provided in terms of policies for today, not for 100 years from now? The real value of the project becomes manifest here.

A fundamental point is that whether the sea is going to rise 3, 5, or 7 ft is unimportant. The important point is that this study indicates a sense of direction that has absolute relevance today.

For example, my office is now conducting a hurricane mitigation study for FEMA. One thing Chapter 1 briefly mentions that we did not even consider is that, with the warming of the ocean's surface, the center of hurricanes may gravitate northward and be closer to Long Island. Furthermore, an 8 ft sea level rise would destroy more than $3 billion of industrial and commercial activities there each year. That includes not only fisheries and tourist industries but all the related industries that go with them. And the homes of almost 1.2 million people would be affected. Therefore, the kind of research presented here today forces me to conclude that this work must continue—particularly the meteorological work.

Other policy implications should jump right out at the federal establishment and, incidentally, should be very popular with the current administration. For example, we have been trying to establish a National Seashore on Fire Island. Since 1964 we have had to deal with New York City people who want to live on Fire Island. They exert their political pressure to keep the Department of Interior from buying up the lands that were originally provided for in federal legislation. The federal government further compounds the problem by issuing insurance to people whose homes may float out to sea. (I have been fighting against flood insurance for years.) In March 1983, five more homes on West Hampton Beach were destroyed by high waters and winds. I cannot say it is because of the rise of the sea; that is a legal question. The citizens who lost their houses said it was flood related and want to get paid. FEMA says it may be an erosion problem. The small print says FEMA will pay if your home is flooded; but they do not pay if erosion causes the loss. I believe these houses were the victims of erosion due to an uncompleted engineering process that started in the 1960s. Suffolk County built 16 out of a planned 21 groin field. Political pressures stopped the project. Of course, the interruption of sand flow downdrift from the sixteenth groin increased erosion.

At Moriches, the barrier beach was broken through. The Corps of Engineers generously helped us fill the breach with a $14 million grant. At the last count, about $5 million worth of that sand has washed back out to sea again. In terms of mitigation for the barrier beach, the latest figure for beach nourishment is $44 million. That does not bother us because most of that is going to be picked up by the Corps. What bothers us is that the county has to maintain the beach at a cost of $8 million every two years. If we can get the Corps to pick up the maintenance costs, we will go ahead with almost any kind of project they wish.

But that policy is not what leads me to my last point. One important area of research is missing in the current study—namely, planning options as an alternative to structural solutions.

The issues presented here have nonstructural solutions. I grant that they are often soft and mushy, but in terms of costs and benefits I think we can benefit from how EPA treats general wastewater management. Looking at the parallel range of nonstructural solutions will provide the full scenario of options in terms of answering questions such as: do you mitigate and stay where you are or, do you take planning steps that allow natural change to take place? This last area of research may fulfill the need for comprehensive general policies. In this fashion we can at least hitch our ride in the helicopter.

As a final comment, I would note that I am very impressed with the harmony and interdisciplinary play among economists, engineers, geologists, geomorphologists, hydrologists, and so on. This is nothing short of remarkable.

Index

Contributors

Michael C. Barth is a principal of ICF Incorporated. He holds a Ph.D. in economics from the City University of New York.

Jeffrey R. Benoit is a coastal geologist with the State of Massachusetts Coastal Zone Management Program. He holds an M.S. in marine geology from the Georgia Institute of Technology.

Charles E. Fraser is chairman emeritus of Sea Pines Corporation. He holds a J.D. from Yale University Law School.

Timothy J. Flynn, a former research assistant with ICF Incorporated, holds a B.S. in Environmental Resource Management from Pennsylvania State University. He is currently pursuing graduate work in hydrology at the University of Arizona at Tucson.

Sherwood Gagliano is president of Coastal Environments, Inc. He holds a Ph.D. in geography from Louisiana State University.

Michael J. Gibbs is a senior associate at ICF Incorporated. He holds an M.P.P. from the John F. Kennedy School of Government, Harvard University.

James E. Hansen is a physicist and head of the Goddard Institute for Space Studies (GISS). He holds a Ph.D. from the University of Iowa.

Miles O. Hayes is a coastal geomorphologist and president of Research Planning Institute. He holds a Ph.D. in geology with support work in marine science from the University of Texas.

John S. Hoffman is director of the U.S. Environmental Protection Agency's Strategic Studies Staff. He holds an M.C.P. from Massachusetts Institute of Technology.

John R. Jensen is associate professor in the Department of Geography at the University of South Carolina. He holds a Ph.D. in geography with a specialization in remote sensing and air photo interpretation from the University of California, Los Angeles.

Timothy W. Kana is director of the Coastal Dynamics Division of Research Planning Institute. He holds a Ph.D. in beach dynamics/ sediment transport from the University of South Carolina.

Murray Kenney is an associate at ICF Incorporated. He holds an M.A. in economics from the University of Sussex, England, where he was a Marshall Scholar.

Lee Koppleman is executive director, Long Island Regional Planning Board. He holds a Ph.D. in regional planning from New York University and currently holds a professorship in urban planning and environmental studies at the State University of New York at Stonybrook.

Andrew A. Lacis is a physicist at the Goddard Institute for Space Studies, specializing in atmospheric radiation. He holds a Ph.D. from the University of Iowa.

Stephen Leatherman is associate professor of geography at the University of Maryland. He holds a Ph.D. in environmental sciences from the University of Virginia.

Gerard P. Lennon is an associate professor of civil engineering at Lehigh University, where he specializes in groundwater hydrology, containment transport, and coastal processes. He holds a Ph.D. from Cornell University.

Colonel Thomas H. Magness, III, formerly assistant director of civil works for environmental affairs of the U.S. Army Corps of Engineers, is science and technology adviser for the Council on Environmental Quality. He holds a B.S. from West Point.

Jacqueline Michel is project manager and director of geological sciences at Research Planning Institute. She holds a Ph.D. in geochemistry from the University of South Carolina.

David H. Rind is a meteorologist at the Goddard Institute for Space Studies. He holds a Ph.D. in geology from Columbia University.

Gary L. Russell specializes in global climate modeling at the Goddard Institute for Space Studies. He holds a Ph.D. in mathematics from Columbia University.

Edward J. Schmeltz, P.E., is assistant vice president and department manager for coastal engineering at PRC Engineering. He holds an M.S. in civil engineering from Texas A&M University.

Robert M. Sorensen is professor of civil engineering at Lehigh University, where he specializes in coastal hydrodynamics, wave/structure interaction, and coastal structure design. He holds a Ph.D. from the University of California at Berkeley.

James G. Titus, of the U.S. Environmental Protection Agency's Strategic Studies Staff, is EPA's project manager for sea level rise. He holds a B.A. in economics from the University of Maryland.

Stuart G. Walesh, P.E., is an associate of Donohue & Associates, where he is responsible for water resources engineering projects. He holds a Ph.D. in civil engineering from the University of Wisconsin at Madison.

Richard N. Weisman is associate professor of civil engineering at Lehigh University, where he specializes in surface water hydrology and coastal processes. He holds a Ph.D. from Cornell University.